T0292712

Managing the Experience of Hearing Loss in Britain, 1830–1930

Graeme Gooday · Karen Sayer

Managing the Experience of Hearing Loss in Britain, 1830–1930

palgrave
macmillan

Graeme Gooday
School of Philosophy, Religion and History of Science
University of Leeds
Leeds, UK

Karen Sayer
School of Arts and Communication
Leeds Trinity University
Leeds, UK

ISBN 978-1-137-40687-3 ISBN 978-1-137-40686-6 (eBook)
DOI 10.1057/978-1-137-40686-6

Library of Congress Control Number: 2017944558

Printed on acid-free paper

This Palgrave Macmillan imprint is published by Springer Nature
The registered company is Macmillan Publishers Ltd.
The registered company address is: The Campus, 4 Crinan Street, London, N1 9XW, United Kingdom

To all those who have experienced hearing loss now or in the past

Foreword

We have so much to learn from history, and a full appreciation of where we have come from can have a profound impact on how we operate in the here and now. Hearing is a pressing contemporary concern with 11 million people in the UK alone having some degree of hearing loss, but of course people experiencing hearing loss have been around for as long as humankind. However, hitherto we have only had histories from the viewpoint of either the (fully) hearing or accounts either by or about people who are profoundly deaf.

The co-authors are changing that with this book, telling the stories of those in the UK who experienced hearing loss before they had access to the care of the NHS, which was only created in 1948. I'm particularly excited by the book as it aligns so closely to the aims of Action on Hearing Loss to change public attitudes, so people better understand the experiences of hard of hearing people and particularly to show how such people have changed the world for the better. There were both hard of hearing Victorians like Harriet Martineau and George Frankland who wrote to explain their experiences and to persuade others to challenge those who exploited their condition, and there were organisations like the National Institute for the Deaf (now Action on Hearing Loss) that worked to stop exploitation of the hard of hearing by sellers of fraudulent 'cures for deafness' and fake hearing aids.

We see how the relationship between medicine and people with hearing loss has been changed too—from doctors' attempts to cure or prevent deafness, to a more supportive approach that became modern audiology. Even so, the book shows how hard of hearing people have faced tough choices about how to communicate, as using hearing aids or lip-reading techniques has always been relatively poorer in terms of giving assistance than, for example, glasses are to someone with visual impairment. These practical challenges have been made worse by public attitudes throughout history, sometimes in very stark ways such as the early twentieth-century eugenics campaigns to eliminate deaf people.

For any and all those who face any such challenges today—or whose loved ones do—this book shows many have felt isolated and frustrated by hearing loss, even bereaved at the loss of easy social conversation. You are not alone, and this book shows how many have found a way, when properly supported to manage a fulfilling life with hearing loss, even if very few of them have left us clues to know how they did it. Few as those stories are, I am delighted that the Action on Hearing Loss Library has provided much of the material for this book and can provide the resources for much future research on hard of hearing people.

This book is part of an ongoing, living story, and I know the co-authors hope (as do I) that another volume will bring the story up to date to show how even with the NHS, hearing loss remained a challenge for many—only now starting to lose its stigma.

I hope that in reading this book, you will not only learn much about a story that has badly needed telling for a long time, but you will also be inspired to find out about the work that Action on Hearing Loss is doing today to help people confronting deafness, tinnitus and hearing loss to take control of their lives so they can live the life they choose.

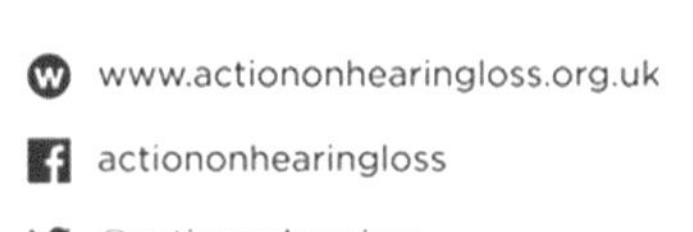

March 2017

Paul Breckell
Chief Executive
Action on Hearing Loss
Royal National Throat, Nose and Ear Hospital
Grays Inn Road, London

Preface

This book began with a request from the Thackray Medical Museum to help with the interpretation of their hearing aid collection, a collection that came largely from the British Society of Audiologists (supplemented by materials from manufacturers) and is arguably the largest in the UK. However, we quickly realised that focusing on hearing aids, while revealing something about the lives of the hard of hearing, failed to encompass other ways in which such people addressed what they experienced as hearing 'loss'—or the subsequent effects and experiences of that hearing loss as adults. It also became clear, thanks to the work of historians of Deaf people and the Deaf comedian John Smith, that hearing aids acted as markers of hearing lives (as assistive technology for the hearing, not the Deaf): as Smith puts it 'hearing aids are for the hearing'. Our book is therefore about hearing people, not the Deaf.

Many seminar and conference attendees offered us their own experiences of many types and degrees of experience of hearing loss and told us that this was the first time that many of them had been able to talk about it in a sympathetic environment. From these conversations, it became increasingly clear that the sense of 'loss' (the sense of grief associated with the loss of hearing for its own sake, but also for lost ease of communication and connection) to them as hearing people was key. This was reinforced for us both through the personal experience of living with and later supporting parents who had lost part of their hearing at various points in their lives, and through the sense of anxiety experienced in being told that one has already lost some of one's own hearing.

We chose the period 1830–1930 because the growth in the world of the technological offers for hearing loss, changes in medical interventions, and cultural changes in attitudes to the hard of hearing constituted a phenomenon worthy of historical explanation.

Leeds, UK

Graeme Gooday
Karen Sayer

ACKNOWLEDGEMENTS

The co-authors would like to thank the following for their invaluable assistance and inspiration in preparing this book: Mara Mills, Jaipreet Virdi, Coreen McGuire, Sean McNally, Action on Hearing Loss, Dominic Stiles, Mike Martin, Stuart Murray, Nick Thyer, Mike Gulliver, Mike Mantin, Myk Briggs, Hannah Hunt, Iain Hutchison, Iwan Morus, John Harley Warner, Jill Jones (DEX), Julie Anderson, Joanne Bartholomew; Lauren Ryall-Stockton, Alan Humphries, Jamie Stark, Alinka Greasley, Janet Greenlees, Annie Jameson, Margaret King, Neil Pemberton, and our late parents. The key phrase 'Hearing aids are for the hearing' comes from the stand-up performance of John Smith, Deaf Comedian, Beautiful BSL, at the conference Disability & the Victorians: Confronting Legacies, Leeds Trinity University, August 2012.

Parts of this book have been presented at the following conferences: Society for the Social History of Medicine conference 2012, 2014, and 2016; British Society for the History of Science 2013 and 2015; History of Science Society 2013; Leeds Centre for Victorian Studies 'Disability and the Victorians: Confronting Legacies' Leeds Trinity University 2012; the Three Societies BSHS/CSHPS/HSS meeting in 2012, and in research seminars at the University of Warwick, University of Leeds, Liverpool Hope University, Leeds Trinity University, University of Oxford, Yale University, University of Cambridge, and Edinburgh University. We are grateful to all in the audiences who supplied constructive feedback on our presentations, helping thereby to make this book immeasurably better than it might otherwise have been.

Finally, we are indebted to the libraries of the University of Leeds, Leeds Trinity University, Action on Hearing Loss and the Thackray Medical Museum in Leeds for access to the invaluable resources in their collections.

Contents

Abbreviations

NHS	National Health Service
NID	National Institute for the Deaf

List of Figures

Epigraph

[To the Editor of *The Lancet*]

Sir,

It has been suggested to me by more than one medical man that I should record the experience of being completely deaf. In my case the loss of hearing was gradual, accompanied for a long time by most severe attacks of vertigo, amounting at last to fits, and cured only by an aural operation.

I have now been quite deaf for about two years. What is the sensation? There is a sense of blankness. I feel always in a condition of expecting to hear, yet never realising the expectation. I cannot describe it in any other way. I have often observed that deaf people have a surprised look about them, as though they were wondering why they cannot hear, and sometimes I wonder if I am getting the same look, for it is not becoming. Then, together with this sense of blankness is the ever-present sense of loss. This at times is overwhelming. I cannot now remain in a room where music is being performed, as the fact that I cannot hear it and never shall is too much for me. For the same reason my duties as a clergyman have become a real and heavy burden, and most of the joy I once had in them has gone. This doubtless is the cause of the depression com mon to most deaf people, whereas the blind who possess the faculty of hearing are usually cheerful.

But the loss of hearing is undoubtedly felt more from its social side. Neither lip reading nor sign language can compensate for the loss of the human voice. We cannot join freely in the conversation of the family circle; and the funny story ceases to be really humorous when interpreted by signs. Rightly or wrongly, this encourages the tendency to withdraw from society and social life. You cannot hear, it is most difficult to join in the conversation, it seems only waste of time to remain, and the time one feels would be better spent in reading or writing.

One of the most distressing results of this total deafness is, to me at least, the horror of the awful silence in the dark, and I should like to know how far this is the experience of others similarly situated... If there are any compensations I have not yet found them, except perhaps that one can read and write undisturbed in a noisy room, and that one's meditations are not disturbed in the London streets or the interior of a motor-bus.

near Rochester F.J. Hammond
January, 1911[1] All Hallows' Vicarage

Note

1. Letter by F.J Hammond. All Hallows Vicarage, nr Rochester quoted in MacLeod Yearsley, 'The Experiences of Deafness' *The Lancet* Feb 4, 1911, 330–331. For the previous medical interventions by Rev Hammond's physician, see Percival Macleod Yearsley, 'A Case Of Severe Vertigo And Tinnitus; Destruction Of The Labyrinth; Cure'. *The Lancet* 172 (1908), 871–872; [Continued] *The Lancet* 173 (1909) 1779.

CHAPTER 1

Introduction

Abstract For the century from 1830 to 1930, we look at hearing loss as a bereavement in human communications, and at the evidential difficulties of recovering the emotional experiences of even the most famous hard of hearing people. A major running theme is the changing prerogative of the medical professions, especially looking at how the role of the otologist evolved, moving away from aural surgeons' attempts to 'cure' deafness to a more adaptive role that sought to assist hard of hearing adults with life's practicalities.

Keywords Communications · Deafness vs hearing loss-changing medical prerogatives

One in six people in the UK during their lifetime experience some form of hearing loss—a figure that apparently remains stable over time and is mirrored worldwide.[1] This book is about the adult population of the United Kingdom who experienced this kind of loss before the advent of the National Health Service (NHS) in 1948. Specifically, it is about those who started life with some hearing and continued to self-identify with 'hearing' culture[2] well after experiencing their loss: rather than identifying themselves as 'Deaf', many sought to 'pass' as hearing, often in the face of ever greater challenges to quality of life from chronically decreasing hearing capacity (often due to more than one cause) during their lives.

G. Gooday and K. Sayer, *Managing the Experience of Hearing Loss in Britain, 1830–1930*, DOI 10.1057/978-1-137-40686-6_1

The different reasons for experiencing hearing loss included: injury, illness, ageing with or without inheritance factors, and some such as tinnitus ('ringing in the ears') with as yet undetermined causes. Not all forms of hearing loss were treated with equal sympathy, and this, intersecting with social class, gender and other factors, had differentiated material effects upon employability, economic security, and access to social, cultural and (pro)creative choices. For example, survivors of the Great War trenches who were deafened by exploding shells experienced hearing loss rather more traumatically than those for whom early onset acquired deafness was a characteristic inherited family trait. We show how their loss of hearing led the war-deafened to receive more sympathetic treatment than those who were deafened by illness or non-combat injuries.

In writing about the history of hearing loss our aim is to recapture how adults managing a fading or vanished capacity for conversation had to deal with others who aimed to mitigate and or profit from what was often their feeling of sensory bereavement. As illustrated by the epigraph of the Reverend Hammond in 1911, the predominant emotions were of isolation and desolation, especially when those who remained hearing responded to them with irritation. Not all 'deaf people' lost their hearing as extensively as Rev. Hammond, and their experience of loss was accordingly less acute. While some adjusted to this new mode of life by lip-reading or using hearing aids to enhance residual hearing, these techniques did not fully restore the prior experience of hearing. For those who identified as hearing, verbal communication was rarely returned to its previous condition, and considerable effort was thus generally required to sustain everyday life in the hearing world.

As the old term would have it, these people were 'hard of hearing': the qualifier 'hard of' meant 'not easily able to'. In pre-modern times this locution referred much more broadly to a whole spectrum of challenging conditions such as being 'hard of understanding', 'hard of believing' and 'hard of knowing'. These are now obsolete, and this form of language is only otherwise used by extension satirically to the 'hard of listening' signalling those who decline to make the 'effort' to listen. And except in a satirical sense those who are partially sighted are not conventionally described as 'hard of seeing'.[3] The difference perhaps consists in the labour involved: the use of hearing aids has generally required far more skill, patience and effort than the successful wearing of glasses. Indeed, eye spectacles are rarely abandoned into a drawer as unusable after just one day of use, as was so often the fate of many an early electronic hearing aid in the NHS system.[4] Hearing and listening, once hearing had been 'lost', could be exhausting, and the assistive technology not worth the effort.

Yet if we want to understand how hearing loss became more greatly stigmatised than loss of vision, we should not look to the amount of compensating labour involved. As we shall see in this book, the differences of status stem more from the much closer cultural associations between deafness and genetic/intellectual 'deficiency' than for short-sightedness. If anything the wearing of spectacles has been more associated with elite literary cultures that were far from being marginalised or stigmatised, unless linked to the comedy of tripping and falling: class and the need/otherwise for physical over literary accomplishment was then the deciding factor. This contrast between the status of sight loss and hearing loss will be a touchstone of our book as we explore the resiliently greater sympathy for lost sight than lost hearing, despite the fact that both those with sight loss and those with hearing loss were subject to new forms of testing and technocratic 'correction' in the West during the nineteenth century.[5] Tellingly, there has only been occasional recognition that hearing loss has the potentially much more alienating and isolating effects outlined in the epigraph, an experience exacerbated by the comedy commonly extracted from the communicational mishaps of acquired deafness.[6] While the comparison of blindness and deafness is a common trope throughout the period we study, notably for Helen Keller who experienced both, and even before,[7] it is a different binary relationship that underpins the stories in this book. For the older adults that feature in our book, sight loss and hearing loss were part of the mutually entangled processes of aging.

1 Hearing Loss History vs. Deaf History

A further key historical entanglement we contend with in writing this book is that in the nineteenth century both acoustic 'loss of hearing' and Deafness as expressed through sign language usage were commonly covered under the rubric of 'deafness', and arguably so up to the mid-twentieth century. The term 'deafness' referred broadly to many kinds of experience, ranging from those who encountered hearing loss only in later life to those born deaf, each of these in varying type (mild to profound, differing by pitch and decibel, unilateral and bilateral): from the point of view of nineteenth century hearing culture, this was construed as a continuous spectrum of hearing. In 1889 the Royal Commission on the 'blind, the deaf and dumb' concluded that 'if we could classify the deafness of the whole population we should find a complete gradation from perfect hearing down to no hearing at all.'[8] This reflected the Commission's view of full hearing capacity as the norm from which deviation occurred by ageing, disease or injury. From the point of view of twenty first century Deaf culture, however, the normalised reference point is not auditory, but the capacity for using sign language, in which only a small minority of the population has ever been proficient.[9]

Our work owes much to historians of Deaf culture, whose work on sign language establishes a baseline of culturally stable 'Deaf' identity. By contrast, we narrate the shifting diachronic process of hearing loss.[10] And our work intersects with Deaf history where we look at past heterogeneous communities broadly identified in the nineteenth century sense as 'deaf', with alliances between partially hearing people and sign-language users that have recently been recaptured in pluralist interpretations of multiple (and often intersecting) 'Deaf Identities'.[11] Amongst the variety of deaf identities that historians can now recover, one visual clue for our study stands out as an element of ordinary, everyday life. Hard of hearing people often used the 'cupped' hand both as a form of manual hearing assistance, and as a visible invitation for interlocutors to adapt their communications accordingly. Thus at the front of Martineau's posthumous autobiography of 1877 there is an unsigned portrait of her, dating 1833, in which a very youthful Martineau cups her right hand to her right ear and awaits conversation[12] (insert illustration here from Maria Chapman Weston, *Harriet Martineau's Autobiography vol.* 1 1877—from Graeme Gooday's personal collection) (Fig. 1).

But, the legacy of the cupped hand as a signal of hearing loss is now visible only as an artistic trope of hearing loss in past visual depictions of eminent deafened subjects such as Thomas Edison, William Gladstone,[13] Harriet Martineau and Joshua Reynolds.[14] As the evidence of the cupped hand falls away, so we are left to infer more about the lives of hard of hearing people from the hearing aids that they used. When we look at older more visible surviving historic devices in museums, these often show signs of modification and wear and tear by their users which tell us much about how much they were regularly used, and perhaps even cherished.[15]

The techniques used by hard of hearing people in the pre-digital era (hearing aids and lip-reading) have been explicitly rejected in the Deaf history of the last few decades; that history has focused instead on the

Fig. 1 Illustration, showing cupped hand, from frontispiece of Maria Chapman Weston, *Harriet Martineau's Autobiography vol.* 1 1877

story of the late-nineteenth century repression and then flourishing of sign language in Deaf culture since the 1960s.[16] Deafness is political because of the very real consequences that have accrued around the social significance of hearing in our period: the hearing are not named as such, yet can determine the outcomes. This rejection of hearing aids as belonging to hearing culture by Deaf culture has been mirrored by an indifference to hearing aids among conventional historians of hearing cultures who have tended to assume a widespread normalised hearing capacity.[17] Hearing loss has a history of vanishing between these two dominant narratives, yet has been part of the process of determining the boundary between Deaf/hearing, and the role of hearing aids as part of a technocracy of normalising hearing has as a result been overlooked. In fact, historic assistive devices for hearing have, until relatively recently, been treated largely as bygones for the museum if noticed at all by mainstream historians.[18] This invisibility has been further reinforced in advertising where there has been a presumptive miniaturization of electronic hearing aids to a discreet near invisibility, entangled with an equally presumptive narrative of 'progress' as hiding deafness from public view.

2 Surveying the Literature

While sound studies[19] has recently emerged as a historical facet of science & technology studies, little account is given in this genre of non-normative hearing. For example, Jonathan Sterne, in *The Audible*

Past: cultural origins of sound reproduction (2003) argued that between the mid-eighteenth and early twentieth centuries, sound itself became an 'object and a domain of thought and practice' while hearing was apparently 'reconstructed as a physiological process'. In his account all people are presumed able to use full hearing capacity techniques to harness, modify and shape their powers of auditory perception in the service of a mass-produced and industrialized 'rationality.'[20] In this domain, there is neither space for variant capacities nor any hint of a rejection of the hearing mode in favour of other sensory engagements altogether.

By contrast curatorial and museological histories, linked to audiological histories, tend to problematize hearing loss as a phenomenon as if it has always been in need of a technological 'solution'. For audiologist turned historian, Keith Berger the story of hearing aids is a linear narrative of increased powers of amplification, moving from shells and horns, to microphone, valve and then transistor-based devices.[21] Similarly for oralist deaf educator, Ruth Bender, the 'conquest of deafness' has been accomplished by ever more systematic use of hearing aids and lip-reading.[22]

More positive approaches to our topic are apparent from Cultural history and Victorian studies: especially material culture and histories of soundscapes. However, again the focus in these genres is on the experience of those who can hear, not on those with variant hearing capacities. The key point is that, these approaches recognise that hearing aids were 'things' that circulated in everyday life and contributed to social status, subject to the characteristic forms of relationship between designers, users, and user-designers. By revealing some of those relationships, the material presence of hard of hearing in Victorian life becomes more tangible.[23]

Similarly helpful is the history of emotions, often allied with the history of medicine, for this specialist sub-discipline points us to ways of enabling exploration of the affective culture of hearing loss. This is crucial for us to flesh out the sense of sensory 'loss' in acquired deafness as a kind of culturally embedded mourning, albeit visible only to those who had lost their hearing. The attempt to reverse this loss was a real and lasting historical phenomenon that drove many to seek mitigating assistance from hearing aids, letter writing and lip-reading. As Disabilities Studies scholar Michael Davidson has argued, using his experience of sudden hearing loss, the 'affective realms of frustration, loss, and failure … are seldom acknowledged experiences of deaf and hard-of-hearing persons', yet are crucial.[24]

As mentioned above, Disability history has not conventionally treated the hard of hearing as part of its remit, and the category 'disabled' is refused in Deaf history.[25] Recently, however, three historians have looked at the experiences of a broader range of deaf people in relation to both their encounters with the medical profession and the technocracies of hearings. Jaipreet Virdi has investigated the hazardous medical encounters of Deaf subjects with aural surgeons in nineteenth century Britain. Jennifer Esmail has noted the presence of hearing aids on the fringes of Deaf sign language culture in the same context, while Mara Mills has explored the complex role of telecommunications on the rise of hearing technologies in the twentieth century USA.[26] Our study complements their work by bringing the past lives of hard of hearing people to the fore in the UK. We explore how lip-reading, writing and various kinds of hearing aids served as complementary resources for the hard of hearing. Also we examine how the medical professions concurrently reconfigured their role in moving from attempts to cure deafness by surgery, to attempts to prevent deafness occurring and to mitigating the effects of acquired hearing loss in recommending hearing aids and lip-reading.

3 Key Themes

In many ways the daily management of hearing loss was a practical matter of accomplishing communications in different environments (e.g. illuminated and acoustic contexts). This could be a matter of using a hearing aid, lip-reading, writing letters, scribbled notes or sign language. But this was no simple unconstrained choice. Adopting a hearing aid required the financial resources to pay for a device and the time and patience to get used to working with it so that conversation could become intelligible. Lip-reading required intensive training and favourable lighting conditions, as well as a limited number of fellow-conversationalists speaking with well-defined, but not over-exaggerated, lip-movements. The use of letter writing techniques was premised on access to a network of like-minded and like-educated correspondents that had somehow been assembled from much dispersed geographical if similar social origins.

To those wishing interlocutors to write down their otherwise inaudible thoughts on a scrap of notepaper, there was much reliance on the patience of conversationalists, their literacy and their willingness to engage in asymmetrical discussions. Finally, for those wishing to use sign language (in its many variant forms), this required access not only to training but a community which used this language. None of these were obvious or easy choices, and the constraints of finances, limited opportunities and spatially configured communities made decisions difficult. While physicians, hearing aid vendors, handbook writers, mail order companies and newspaper advertisers all had much advice to offer, it came at the cost of either a service provided or a dependency relationship created.

One key question is 'what was it like to be deaf in this period?' But recovering the testimony and agendas of the hard of hearing is very difficult given how little evidence is left behind of how they made such choices. More often we learn of the experience of hearing loss from the other—from those who found it frustrating, or sought to study it as a research topic, or who sought to profit from the solutions. As with others who experienced disabilities or poverty, most written evidence tells us about others' attitudes to those thus affected, and tend to be accusatory, or condescending rather than their examining the experiences of hearing loss directly. Where we have direct evidence of the views of hard of hearing people it is from three kinds of course: (1) either solicited/mediated by physicians—as for the Rev. Hammond above (2) from journalistic accounts from George Frankland in exposing fraudulent vendors of hearing aids, or (3) letters to magazines or newspapers from those seeking to make others aware of their condition, often pseudonymous such as 'Out in the Cold'.

Whereas Harriet Martineau (1802–1876) was a very well-known and explicitly self-identifying 'deaf' person who talked much about her loss of hearing, there is a cultural invisibility of acquired hearing loss/deafness, even among the notable individuals of the time, such as British Prime Minister William Gladstone (1809–1898)[27]; British electrical engineer Ambrose Fleming (1849–1945) prize-winning electrical writer Oliver Heaviside (1850–1925); Princess Alexandra of Denmark/Queen Alexandra of Great Britain (1844–1925)[28] who married the Prince of Wales (later, Edward VII) in 1863; Queen Victoria, who acquired a hearing device c.1880 (aged c.61), but could also communicate before that in sign/fingerspelling[29]; and Winston Churchill (1874–1965) who experienced late-life hearing loss like his father.[30] While the socially confident hard of hearing like Harriet Martineau were determined to be publicly explicit about their deafness, not all felt a wish to do so. Others sought to 'pass' as hearing by using suitably disguised or discreet hearing aids or invisible lip-reading so that none need know of what was too often felt by those concerned to be an embarrassing situation, even a deficiency.[31]

Finally, as a thematic point, we should consider broader cultural factors which changed this situation. One key point was the position of the medical profession: by the mid-twentieth century it had abandoned the conceit of attempting to 'cure' all forms of deafness by surgical means. Once the discipline of otology had emerged in the early twentieth century it was a field more aligned to the prevention of deafness and treatment of specific aural diseases. Only with the advent of the First World War did the medical profession find itself in a new position. Asked to adjudicate on cases of hearing loss caused by explosion-prone life in the trenches, otologists found themselves not only the arbiters of who had lost hearing and how much; they were also asked to comment on the financial compensation or other support that should be granted to deafened servicemen (mostly men anyway). As explained in the epilogue, this constituted both part of the pre-NHS framework of state obligation to the war-injured and also the origins of what later became known as audiology.

But how did physicians learn to accept that their clinical expertise in aural surgery was insufficient to the task, and they had instead to appropriate resources from elsewhere, notably advice on hearing aids and lip-reading? That is the explanatory challenge for this book. A helpful broad perspective can be found in a lecture of the retired Glaswegian otologist James Kerr Love in 1936 in 'Deafness–Its Prevention and Relief' presented to a group of postgraduate medical students. From several decades in the business, he had come to realise that there were limits in the capacity of his profession, and also strategic ways of allocating assistance for adult deafened people—an activity in which he was much involved in the Glasgow Hard of Hearing Club, the first of its kind in the UK set up in the mid-1920s.

> Seeing we must face the facts that the deaf and the hard of hearing, like the poor must continue to be with us, that neither surgery nor medicine will cure chronic deafness, what can be done for these severely handicapped people? Deaf clubs for the hard of hearing are based on the understanding that treatment has been tried and has failed.

Kerr Love's language of the 'handicapped' is illustrative of a lingering eugenic interpretation of deaf people as somehow disfavoured in life. And conceding the failure of the medical curative project for chronic conditions, his conclusions about non-surgical interventions were very clear: lip-reading was really best given to younger hard of hearing people; for those at middle age and over a hearing aid was required:

> But the phrase "hard of hearing" implies that in most cases some hearing is left. One of my old teachers used to twit me about the inability of the aurists to cure chronic deafness. But can we cure chronic anything - chronic Bright's heart disease for instance? The answer to the question what can be done? is- If the subject is young teach him lip reading. This hardly ever fails with a young person with good eyesight. If he is at or past middle life give him an "aid" to his remaining hearing. Lip reading is the best "aid" we have and, except in the dark, is always available. No apparatus need be carried about by the good lip-reader.[32]

This was not necessarily the view of all otologists in the 1930s: some long maintained the view that hearing aids permanently damaged hearing, and were thus to be avoided.[33] Yet we need to explain how this major transformation had occurred: the prescription of hearing aids and lip-reading were not previously any direct part of medicine: how did they become so? And then how did other organisations such as the National Bureau for the Encouragement of the Welfare of the Deaf in 1911 come to be reconfigured as the National Institute for the Deaf (NID) in 1924, fighting for the interests of the hard of hearing and defending them against opportunist vendors. As we shall see a key part of the story was the First World War and its prompting of the government to intervene.

4 The Structure of the Book

Overall this volume covers a century from c.1830 to c.1930. We start in the 1830s when medical 'cures' for deafness were really significantly challenged by recommendations to use hearing assistive devices from both an aural surgeon, John Harrison Curtis, and a self-styled 'deaf' writer Harriet Martineau. We end in the 1930s when the National Institute for the Deaf acted to represent the interests of all kinds of deaf people in managing relations with hearing aid vendors, a new generation of otologists had replaced aural surgeons in systematically recommending hearing aids and or lip reading for the hard of hearing. How that transformation in prerogatives and identities was accomplished over that century is the major task of this book, with Harriet Martineau's 'Letter to the Deaf' of 1834 serving as a touchstone for how discussions moved in favour of the hard of hearing away from a fully medical control of deafness.

While there are strong chronological elements to our story, our approach to organising each chapter is more fundamentally thematic than chronological. We move from this Introductory chapter to Chap. 2 *'The diverse and changing categories of deafness'*. This shows how the very broad nineteenth century umbrella notion of deafness covered various different kinds of experience, among which being hard of hearing was just the most pervasive. We explore how these categories developed as the understanding of multiple causes of acquired deafness in disease and aging developed further differentiations, including the advent of 'war-deafness' in the terrible conflict of 1914–1918. Chap. 3 examines the various kinds of *Advice for managing hearing loss* starting with the medical advice on curing hearing loss which soon retreated into support for hearing trumpets. We then investigate how Harriet Martineau argued against trust in medical advice and put forward her own example as a self-confident 'deaf' person managing her life with a hearing trumpet. This is complemented by a survey of the often far from disinterested advice from hearing aid vendors about what kinds of hearing trumpet or alternatives which should be purchased.

Chapter 4 *then* addresses the topic of *Communicating with hearing loss* by setting up an experiential comparison between the loss of hearing and the loss of vision to show both how they were inter-related through the reliance of many hard of people on vision to conduct communications. This is explored in relation to both the use of lip-reading, often taught by women, and correspondence networks between deaf people, also characteristically suiting the needs of isolated hard of hearing women. Chap. 5 continues this theme by looking at the *Selling and using of hearing aids* by hard of hearing people that wanted to maintain conversations with the hearing without the awkwardness of reliance on lip-reading or writing. We discuss the many kinds of hearing aid available, disguised or elegantly designed for display, suited variously for many kinds and degrees of hearing loss. Two especially eminent London-based hearing aid companies are discussed: the genteel bespoke personal service of the Rein Company and the more medically-oriented mail order service of Hawskley. The more opportunist hearing aid vendors who sought just to profit from hearing loss are discussed through the eyes of campaigning journalists who were themselves hard of hearing.

Moving towards the early twentieth century Chap. 6 looks at two medical approaches to *Preventing deafness* after discussing the changing cultural climate of fin-de-siècle Britain in which combined eugenic and economic factors threatened the well-being and status of all kinds of deafened people. In this context we discuss two otologists' attempts to (re)medicalise deafness with a new emphasis on how to prevent rather than cure it. We look at Percival MacLeod Yearsley of London and his eugenic obsession with the small minority for which deafness was allegedly heritable and the Glasgow-based James Kerr Love who aimed to show that most acquired deafness arose from health factors relating to poverty. Then Chap. 7 *Institutionally organizing for hearing loss* looks at the organisations that emerged from the late nineteenth century to advocates the rights of deaf and hard of hearing people in response to eugenic and economic discrimination against them. We focus first on the *Deaf and Dumb Times* (later the *British Deaf Times*) as a forum for the heterogeneous deaf community with hard of hearing journalists articulating the nature of their exploitation and repression. Next we look at how the National Institute for the Deaf in 1924 emerged to a new umbrella role: one key factor was the First World War changed the situation as handling the context of combatant hearing loss transformed the advocacy of the pre-War organisations into a more unified national approach. Finally, this brought focused national attention to the needs of hard of hearing people, however they wished to live and communicate.

Notes

1. According to Action on Hearing Loss, one in six people in the UK experiences a disabling degree of hearing loss, especially older groups. http://www.actiononhearingloss.org.uk/your-hearing/about-deafness-and-hearing-loss/statistics/~/media/56697A2C7BE349618D336B41A12B85E3.ashx (last accessed 16th Dec 2016); According to the World Health Organisation over 5% of the world's population has disabling hearing loss and approximately one-third of people over 65 years of age are affected by disabling hearing loss. WHO http://www.who.int/mediacentre/factsheets/fs300/en/ (last accessed 16th Dec 2016).
2. See Rebecca A.R. Edwards 'Teaching Deaf History' *Radical History Review* issue 94 (Winter 2006): 183–190 for a discussion of this in the classroom.
3. See *Oxford English Dictionary* entry on 'Hard' 12c. Hard of hearing, which lists satirical variants in 'hard of listening' and 'hard-of-seeing'.

4. The Medresco system, developed from 1943 by the Post Office and Medical Research Council, was free to the qualifying public from 1948 via the National Health Service—examination via an NHS regional audiological centre and means tested. As Sean McNally's PhD research at the University of Leeds has shown, there was great demand for this.
5. For a discussion of the US case see: Peter John Brownlee, 'Ophthalmology, Popular Physiology, and the Market Revolution in Vision, 1800–1850.' *Journal of the Early Republic* 28, no. 4 (2008): 597–626. http://www.jstor.org/stable/40208136 (last accessed 19th Dec 2016).
6. David Lodge, *Deaf Sentence* (Basingstoke: Penguin, 2009).
7. Emily Cockayne. 'Experiences of the Deaf in Early Modern England.' *The Historical Journal* 46, no. 3 (2003): 493–510. http://www.jstor.org/stable/3133559 (last accessed 19th Dec 2016) 495.
8. *Report Of The Royal Commission On The Blind, The Deaf And Dumb, & C. Of The United Kingdom.* (London: Her Majesty's Stationery Office, 1889) 291. 21.403. This is cited in James Kerr 'School Hygiene, in its Mental, Moral, and Physical Aspects', *Journal of the Royal Statistical Society*, 60, (1897), 613–680. 672.
9. See Jan Branson, Don Miller, *Damned for their difference: the cultural construction of deaf people as "disabled": a sociological history* (Washington University Press: Gallaudet University Press, 2003), 145.
10. Deaf history arose out of the activism of the 1960–1980s around sign language. Much of this work has been led by the USA, particularly by Gallaudet University, where the focus is on those who lost their hearing before 11, and who self-identified as belonging to the Deaf community, especially through the use of sign language (e.g. American Sign Language, ASL). Another centre of focus is France, due to the significance of Abbe de l'Epee, and the influence of French sign language on Thomas Hopkins Gallaudet (via Laurent Clerc)—in founding the first Deaf school in the USA 1817, e.g. C. Aicardi, 'The Analytic Spirit and the Paris Institution for the Deaf-Mutes, 1760–1830' *Hist Sci*, xlvii (2009), 175–222.
11. Irene W. Leigh, *A Lens on Deaf Identities*, (Oxford: Oxford University Press, 2009).
12. Maria Chapman Weston, *Harriet Martineau's Autobiography, vol.* 1 (Boston: James R. Osgood & Co., 1877).
13. In the year of his death (1898), Gladstone was portrayed in his final encounter with Queen Victoria carefully placing his left hand against his ear. The illustration was 'Queen and Premier–The Last Audience' Richard B. Cook, *The Grand Old Man, Or the Right Honourable William Ewart Gladstone four times Prime Minister of England,* (Publisher's

Union, 1898), 'Queen and Premier—The Last Audience,' 42, 523 image available online at https://archive.org/stream/grandoldmanorlif00cookrich (last accessed 22nd Sept 2015).

14. See, for example, the image of the cupped ear in Sir Joshua Reynolds' *Self-Portrait as a Deaf Man* (Tate Gallery). http://www.tate.org.uk/art/artworks/reynolds-self-portrait-as-a-deaf-man-n04505 (last accessed 19th Dec 2016) featured on the front cover of Brian Grant (ed) *The Quiet Ear: Deafness in Literature* (London: André Deutsch, 1987).
15. Graeme Gooday and Karen Sayer, 'Purchase, Use and Adaptation: Interpreting 'Patented' Aids To The Deaf In Victorian Britain' forthcoming in Claire Jones (ed.) *Rethinking Modern Prostheses in Anglo-American Commodity Cultures, 1820–1939* (Manchester: Manchester University Press, 2017), 27–47.
16. The destruction of hearing aids by a hostile Deaf community is illustrated in Edwards, 'Sound and Fury'.
17. Mike Goldsmith *Discord: the story of noise* (Oxford: OUP, 2012) and David Hendy *Noise: a human history of sound and listening* (London: Profile, 2013).
18. Elisabeth Bennion, *Antique hearing devices.* (Vernier Press: London & Brighton, 1994).
19. David Hendy, *Noise: A Human History Of Sound And Listening* (London: Profile Books, 2013); Trevor J. Pinch and Karin Bijsterveld (eds), *The Oxford Handbook Of Sound Studies* (New York & Oxford, Oxford University Press: 2012); Alexandra Hui, *The Psychophysical Ear: Musical Experiments, Experimental Sounds, 1840–1910*, (Cambridge, MA, MIT Press: 2013); Marc Perlman, *Unplayed melodies: Javanese gamelan and the genesis of music theory*, (Berkeley, Calif., London, University of California Press, 2004).
20. Jonathan Sterne, *The Audible Past: cultural origins of sound reproduction* (Duke University Press, Durham & London: 2003) 2–9.
21. Whereas Asa Briggs, *Victorian Things* (Harmondsworth: Penguin, 1990) focuses on the Philosophy of the Eye, there is minimal discussion of the process of hearing or of the ubiquity of hearing aids.
22. Ruth Bender, *The Conquest of Deafness: a history of the long struggle to make possible normal living to those handicapped by lack of normal hearing.* (Cleveland, Ohio: Case Western Reserve University, 1960).
23. See, for example, Jennifer Sattaur, 'Thinking Objectively: an Overview of "Thing Theory"', *Journal of Victorian Literature and Culture*, 40 (2012), 347–357; Alastair Owens, Nigel Jeffries, Karen Wehner and Rupert Featherby, 'Fragments of the Modern City: Material Culture and the Rhythms of Everyday Life in Victorian London', *Journal Of Victorian Culture,* 15:2 (2010), 212–225; Marieke M.A. Hendriksen, 'Consumer

Culture, Self-Prescription, and Status: Nineteenth-Century Medicine Chests in the Royal Navy', *Journal of Victorian Culture*, 20:2 (2015), 147–167.

24. Susan J. Matt and Peter N. Stearns (eds), *Doing Emotions History* (Urbana, Chicago, and Springfield: University of Illinois Press, 2014); Michael Davidson, 'Cleavings: Critical Losses and Deaf Gain' unpublished CCDS seminar paper, Liverpool Hope University, UK Wednesday 18 December 2013.
25. Iain Hutchinson, 'Oralism—a Sign of the Times'. *European Review of History*, 14 (2007), 481–501; see also Kudlick, Catherine J. 'Disability History: Why We Need Another "Other"' *The American Historical Review* 108, no. 3 (2003): 763–793. doi:10.1086/529597; Jan Branson, Don Miller, *Damned for their difference: the cultural construction of deaf people as "disabled": a sociological history* (Washington D.C.: Gallaudet University Press, 2003), 145; Harlan Lane, *The mask of benevolence: disabling the deaf community* (New York: Knopf, 1992).
26. Jaipreet Virdi-Dhesi, 'Curtis's Cephaloscope: deafness and the making of surgical authority in London, 1816–1845', *Bulletin of the History of Medicine*, 87: 3 (2013), 347–377; 'From the Hands of Quacks: Aural Surgery, Deafness, and the Making of a Surgical Specialty in 19th Century London' (PhD thesis, University of Toronto, 2014). Mara Mills, 'Hearing Aids and the History of Electronics Miniaturization', *Annals of the History of Computing, IEEE*, 33:2 (2011), 24–45. Jennifer Esmail, *Reading Victorian Deafness: Signs and Sounds in Victorian Culture* (Athens, Ohio: Ohio University Press, 2013).
27. See letter of thanks from W. E. Gladstone to Burroughs, Wellcome & Co. 1893, Image: M0008363 Credit: Wellcome Library, London, Library reference no.: ICV No 51428, Wellcome Images.
28. There is evidence that she wore an early Akoulallion battery aid from the mid 1890s (the forerunner of the Acousticon hearing aid). See blog discussions by Jaipreet Virdi https://jaivirdi.com/2014/03/25/the-time-travelling-vote-gathering-miraculous-acousticon/ and Neil Bauman http://hearinglosshelp.com/blog/miller-reese-hutchisons-coronation-medal/ (last accessed 20th December 2016).
29. There are several images of her Queen Victorian finger spelling, painted by Deaf artist William Agnew (1846–1914). See Esmail, 2 and H Dominic W Stiles, 'Deaf Artist William Agnew' at http://blogs.ucl.ac.uk/library-rnid/files/2011/12/Agnew-Picture-2.jpg (last accessed 22nd Sept 2015).
30. Martineau and Churchill cases both discussed in Elisabeth Bennion, *Antique hearing devices*, (Vernier Press, London & Brighton: 1994); also see e.g. Michael Hill, Susan Hoecke-Drysdale (eds) *Harriet Martineau*: Theoretical and Methodological Perspectives.

31. Jeffrey A. Brune and Daniel J. Wilson (Eds.) *Disability and Passing: Blurring the Lines of Identity* (Athens, Ohio: Temple University Press, 2013).
32. James Kerr Love, 'Deafness–Its Prevention and Relief', *Postgraduate Medical Journal* 1936, 12: 92–96.
33. See Edwin Stevens, 'Hearing-Aids For Deafness', Letter to the Editor of *The Lancet*, April 30 1938, 1022–1023 criticizing the views of Percival MacLeod Yearsley for maintaining the view that hearing aids harmed the residual hearing of the hard of hearing.

CHAPTER 2

The Diverse and Changing Categories of Deafness

Abstract The very broad nineteenth century umbrella notion of 'deafness' covered various kinds of differential auditory experience, among which being 'hard of hearing' was just the most pervasive. We explore how understandings of deafness developed and multiplied as the manifold causes of acquired deafness in disease and aging developed further differentiations, including the advent of 'war-deafness' in the global conflict of 1914–1918.

Keywords Changing categories of deafness · Identities · War-deafness

When Harriet Martineau published her 'Letter to the Deaf' in 1834, what kind(s) of 'deaf' constituency was she addressing?[1] Her readers might have anticipated from her broad general title, and her self-categorisation as a 'deaf' person that she was speaking to all kinds of 'deaf' people. But in recommending that they share her practice of using hearing trumpets to enhance or restore their conversational capacity it was clear she was only addressing those with some residual hearing. Like Martineau's readers then, we cannot take for granted what any given reference to 'deafness' meant in the nineteenth century. And following Irene Leigh's recent focus on the multiplicity of 'deaf' identities in the Twenty first century, we explore in what follows some of the diverse historical meanings of deafness as they intersect with understandings of hearing loss.[2] Whether Martineau's readers actually identified with her

G. Gooday and K. Sayer, *Managing the Experience of Hearing Loss in Britain, 1830–1930*, DOI 10.1057/978-1-137-40686-6_2

particularised partial notion of 'deafness' is indeed an open question. At the time she was writing there was no tidy taxonomic differentiation of 'hard of hearing' as a gradually acquired mild-severe condition versus 'deafened' as a severe-profound loss. Nor indeed was there a fixed understanding of whether capacity for aural conversation necessarily accompanied any sort of 'deaf' condition as Martineau appeared to presume.

This chapter aims to historicize such issues by showing the contingent and changing nature of the categories of deafness used up to the First World War in all their untidy and non-consensual complexity. We explore how different kinds of deafness are historically and culturally specific in their reference along a rich and continuous spectrum of human experiences. Thus, in what follows we explore first the meanings of hearing, then of 'deafness', e.g. in the common nineteenth-century association of the word 'Deaf' with 'Dumb' for the non-speaking, and 'Stone Deaf' used to refer to complete hearing loss, but without implying any lack of speech.[3] Then we look at the changing categories that arose to modify the spectrum of meanings of deafness in the early twentieth century. This includes the twentieth century eugenic concern to differentiate 'hereditary' deafness and 'acquired' deafness, as well as a new sympathetic concern for those who were extrinsically 'deafened' by the explosive effects of the First World War. The chapter thus closes with a discussion of how early in the twentieth century new experiences of combat-induced deafness brought a new category into existence and a new sympathy and respect for at least some among the broad constituency of the deaf.

1 'Hearing' vs Hearing Loss as a Form of Deafness

In the nineteenth century, as now, the status of 'hearing' vis-a-vis deaf was rarely examined, since in general the hearing population have not been compelled to characterise this normalised condition. Yet a perceived departure from this 'norm' for those experiencing various kinds of deafness becomes clear from the evidence of those who experienced hearing loss. It was characteristically an emotionally isolating phenomenon, as everyday relationships became gradually more challenging for those who could not sustain a tacitly 'normal' performance of hearing. With the possible exception of hearing loss induced in wartime combat, British society/culture has not generally been as sympathetic to the loss of hearing as the loss of sight. Rarely was the onset of hearing loss considered to be a 'tragedy' akin to the trauma of loss of eyesight.[4]

If anything, as we examine later, loss of hearing was popularly represented as somewhere between irritating and comic rather than tragic.

One common stigmatizing trope discussed below was that the supposedly 'deaf' or 'hard of hearing' could in fact hear more effectively than they claimed, but simply *chose* stubbornly not to hear. This unsympathetic representation of the hard of hearing has long been marked by a lack of empathy within the Hearing world which has treated them as 'infirm', for example, weakened by old age, or emasculated for men. The sense of loss experienced was therefore exacerbated by a loss of agency and a sense of inexorable transition into old age and dependency on the assistance of others. To further heighten the sense of isolation, those who became hard of hearing as adults rarely connected with the experiences of the c. 2% born deaf or who became deaf in early life and who managed communication typically by learning to sign without necessarily being aware of any 'lost' capacity for hearing. Hence, although those with a late-onset hearing loss identified themselves as Hearing, they have fallen into a cultural/social limbo between Deaf cultures and Hearing cultures, not clearly belonging to either.

Jennifer Esmail has argued that in seeking to understand Victorian conceptualisations of deafness that 'hearing Victorians' conventionally understood deafness as a pathology, believing that people who experienced complete deafness were suffering under a 'heavy misfortune', and thus presumptively needing 'pity and charity.'[5] As Jan Branson and Don Miller have shown, however, this was not so at the start of the nineteenth century. When charitable and philanthropic voluntary organisations gave non-hearing deaf people religious instruction, they treated them as having ordinary needs while simply being different within a wider community. For the latter part of the nineteenth century Branson and Miller identify a shift in the categorisation which led to this constituency increasingly being pathologised and treated instead as if impaired, i.e. as 'disabled non-hearing people'. They were thus excluded from mainstream society in specialist schools and asylums, without access to conventional education or employment opportunities. Such pejorative cultural constructions of deafness did not just impinge on those who were deaf all through their lives: they would also have had a considerable impact on the life quality of any hearing Victorian who acquired some significant degree of deafness after childhood.[6] For our purposes the risks could be lost prospects of employment or marriageability, as well as social isolation from hearing society.

This alienation from conversational life was often not apparent to those who had no experience of lacking the communicative resource of hearing. The sound world of pre-industrialized societies could be overlaid with meanings that were self-evident only for those with full hearing capacities. For the case of nineteenth century France, Alain Corbin's classic study has explored the history of the meanings attaching to the sound of pealing church bells—warnings, celebrations, calls to prayer or to marriages or funerals. The collective emotional import of a peal of bells could overwrite all other concerns until the advent of secularisation brought the rise of written over auditory forms of authority. Yet, both he and his primary sources take it for granted that everyone could hear the bells and hear them equally. This raises questions not addressed by Corbain concerning how a loss of hearing entailed loss of access to the local soundscape and the consequent loss of informational knowledge and emotionally-fraught sense of displacement.[7]

In the British case to recapture the emotional tenor of that experience of lost access to a sound world, we must turn typically to religious and literary periodicals. These sources are one of the few repositories of narratives of loss of hearing discussed openly by writers who lived through it. After all, to receive some interest or attention from others, those who had lost (some) hearing had to turn their experiences into some culturally aesthetic form easily consumed, whether by deaf or hearing cultures. And it was poems in particular that could concisely convey the emotional tenor of those experiences.[8] Instead of direct complaint, they attempted to establish the poignancy of hearing loss through such recurrent tropes as the increased inability to perceive the sounds of Nature, music, childhood and God/Church. *The Literary Gazette* of 13 July 1850, for example, carried 'The Lament of the Deaf Lady' by Emily Varndell which described a litany of lost sounds: song, the lark 'carolling', church bells (now unable to call her to prayer), the sea, the wind (and, we infer, an Aeolian harp), the bee, the watch dog, a child's lisping, gentle voice. Poetry like this was meant to elicit reflection among the hearing as to their good fortune—thanks due to God that they were not in the subject's position—in order to garner their charitable intervention to assist her. [9]

The Rev. J. Lancaster Ball's poem 'A Deaf Man's Monody in Spring' carried by *The Wesleyan-Methodist Magazine* in May 1886 depicted a man who has lost his hearing and thus access to birdsong, the 'sounds from grove or stream', music, gentle conversation, and 'still small voices'.

This offered a highly evocative and personal account of the experience of hearing loss late in life: 'I still see long-loved faces by/I *see* them speak, but know not what/I see them smile, but know not why/Yet I will be glad and murmur not'. He thus reflected that divine benevolence had left him with the 'greater gift of sight'.[10] Thus we see vision and hearing were placed within a hierarchy with the former ability typically privileged over the capacity to hear. From this we can understand the common—if later contested—nineteenth century perception that loss of sight was a greater tragedy than loss of hearing; by contrast the latter was regrettable, and either irritating or comical depending on context.

Indeed, it was not the poignancy of writers mourning for lost hearing that we find in other kinds of media representations. Instead the hard of hearing were cast much more harshly as irritating in their failure to uphold conversational norms or laughable in the mistakes that they made in attempting so to participate. These alternative tropes fuelled comic depictions in *Punch* and the *Spectator* that characterised hearing loss narrowly as an affliction specifically of bodily deterioration in later life, and one that required a hearing trumpet to remedy. Some such depictions could be strongly gendered: where the hearing trumpet user was a middle-aged male there was a common theme of emasculation in the putative infirmity of deafness. In the satirical mode of these two publications comic effect was created through captioned misunderstandings achieved by mishearing crucial consonants. For example, in the punningly titled 'A Matter of Hearing' that was published by the *Spectator* in 1908, the Witness examined in court declares: 'I'm a bad liar, am I? Then what are you?' to which The Beak (magistrate) replies: 'I asked if you had it on hire?'. Silence in court is then demanded to halt this communicative mishap.[11]

Other cartoons carried in the press sometimes reiterated the trope that the supposedly hard of hearing were being lazy, disingenuous or outright fraudulent: they could actually hear but simply chose not to do so. This supposedly voluntary deafness appeared figuratively in phrases such as 'turning a deaf ear' akin to 'turning a blind eye' in its implications of selective or even irresponsible indifference. This pejorative association was harnessed for political satire by the weekly *Fun* on 9 July 1890 in 'Deaf As a Post: the raikes' [sic] progress *versus* the Post Office pilgrims' progress'. The Postmaster-General Henry Cecil Raikes MP (who served in that office under Lord Salisbury 1886–1891) was depicted with his head atop a post box body, ears and eyes closed to the loud complaints of his employees represented as a single post-man burdened by a huge sack of 'grievances'.[12] Such a representation in a popular, progressive publication could only have reinforced the common lack of sympathy for those who had genuinely lost hearing.

2 Hearing Loss Among the 'Stone Deaf' and the 'Deaf and Dumb'

The derogatory discourse of hearing loss as voluntary deafness can be contrasted to two other subcategories of deafness in the nineteenth century. People who were genuinely and absolutely deaf were known as the 'stone deaf', subsequently re-categorised as the 'Deaf' in the sign-language narratives of the twentieth century. And there were those who did not speak—the so-called 'mute' or 'dumb'. This association was epitomised in one of the titles of the deaf-owned publication *Deaf and Dumb Times* (see Chap. 6); it is significant that this title changed several times owing to the evident dissatisfaction of some of its readers, among other reasons.[13] The way in which all of these forms, along with hearing loss, could be categorised by Victorians under the rubric of 'deafness' illustrates the range of terms and elasticity of the word 'deaf'. This can make it difficult for historians to access the distinct experience of those who considered themselves to have a hearing loss. For example, those categorised as deaf-mute could themselves have previously had hearing and then experienced sudden traumatic loss, as was a supposedly 'deaf-mute' telegrapher in the USA who was reported in 1878 as having lost his hearing suddenly after completing training.[14]

The category of 'deaf and dumb' was used very widely as a bureaucratic phrase in British Parliamentary papers and reports on all kinds of official texts such as the decennial census, which enumerated the deaf in a category with that heading from 1851. For example, consider Boyles' Report to the 'Commission on the Employment of Children, Young Persons, and Women in Agriculture' (1867) which addressed the appropriateness of field labour for girls and women. Of girls who did not enter farm service at the ages of 12–14, but remained employed instead in agricultural labour, a typical observation was that there was 'generally something against them' being either 'deformed or deaf and dumb'.[15] It was thus assumed that the status as 'deaf and dumb' for such young women was akin to (physical) 'deformity' in unfitting them for the socially superior employment of domestic service, so that they were only permitted to do lower status work in the field.

Both this version and an alternative reification of the category of 'deaf and dumb' were utilised in *The British Deaf Mute*, (1895–1896), albeit without fixed protocols and also with separability of terms. This was an activist publication that directly paralleled the circulation of other very active internationalist groups at the time, such as the trades unions and suffrage societies. For example, this publication reported occasionally on such distinctive items as 'Deaf-mute Lady Entertainer', and 'Deaf-Mute Football League' on also on the 'Jews' Deaf and Dumb Home'. Nevertheless, somewhat subverting its own title, reference simply to 'the deaf' (capitalized in the titles of articles) was much more commonplace, e.g. 'The African Deaf', 'Church's Care of the Deaf', 'the New Institute for the Adult Deaf', 'Orally Taught Deaf', 'Schools for the Deaf' and for the 'Deaf and Blind', 'Educators of the Deaf'. On the other hand the conjunction of deaf and dumb were deconstructed in one article titled 'Deaf Children not Necessarily Dumb ("in the Beginning")'. As these headings might suggest, 'deaf', even without 'mute or 'dumb' attached, was rarely used alone. In a publication that followed all things 'deaf', no one it appears was ever purely 'deaf' (whether explicitly or implicitly signalled), and only in relation to the stone-deaf was there an implication of no hearing capacity.

In headings like those just given, or others such as 'the Indian Deaf', 'Pictures by Glasgow Deaf Artists', and 'Proposed Deaf Trades-Union' we see how multiple forms of deaf identity emerged by intersections with gender, religion, class, maturity (child vs. adult), 'race' and or blindness. This suggests that the term 'deaf', far from being associated just with 'dumb' as in the official category, or 'mute' as in the periodical's title, had a wide range of uses and associations among those who experienced any degree of hearing loss, and that the preference for any one of them might vary depending on context even for a single individual. Indeed, working for a British publication interested in promoting unity among the deaf worldwide, the partially-hearing house journalist George Frankland[16] flagged this up quite explicitly in *The British Deaf Mute* in a piece titled 'By What Name?' in November 1895. There was real dissatisfaction with the situation among '[t]he deaf, the dumb, and the deaf-and-dumb or deaf-mutes' who were then 'agitating' for a designation 'comprehending all these classes' since to be described as 'Deaf and

Dumb' was unjust to those who were either but not both. Using explicitly Darwinian language he commented that these were all 'really names of species' for which an overarching genus was needed. This was a difficult matter as the proposed broad category could not be as broad as the 'Infirm' since that could 'include the blind, lame, and idiotic', nor so narrow as 'Deaf' which would 'leave the poor dumb out in the cold'.[17] Accordingly Frankland discussed—perhaps in part, satirically—some possible neologistic alternatives to overwrite all these: 'Silentians' and 'Owrotics' combined from 'aurotics' and 'orotics'.

Such was the difficulty of this point that Frankland felt unable even to start discussion on how far this taxonomical debate could include the concerns of those who were, like him, neither stone-deaf nor mute, but not fully hearing either:

> Possibly many would like neat verbal distinctions between the "Hard of hearing," "Partially deaf," and "Stone deaf," as well as between degrees of dumbness, but I think this question had better be postponed.[18]

Evidently the prospect of unifying all the varieties of the deaf-hearing experience under one banner was a long way off, and Frankland's proposals for an all-encompassing term did not succeed, and he made no further efforts to differentiate between the 'Hard of hearing' and the 'Partially Deaf.'[19]

Soon it was not just the heterogeneous deaf community that was discussing the categorisations of deafness/partial deafness/partial hearing. In the medical domain there were attempts to link the categorisations of deafness to biomedical and environmental causation, decoupled from the lived experience of hearing loss.

3 Medicalising Hearing Loss: Acquired Deafness and War-Deafness

Within a decade of that unresolved debate in *The British Deaf Mute*, the rise of eugenics—a concern promoted particularly by Alexander Graham Bell—posed new threats to the identity and autonomy of people without (full) hearing.[20] The new eugenic quest was to *eliminate* deafness as if an 'undesirable affliction': the causes of deafness would be eliminated to preclude further deleterious economic consequences for the taxpayer who had to support the institutions for their care. This brought into medical forums new questions about the organic and economic categorisations of deafness/hearing loss. The degree or kind of deafness or the subjective experience of hearing loss was of less interest to clinicians who sought to account for the origin of the condition in reductively medical terms. The newly enhanced medical professional interest in framing deafness and hearing loss in terms of the aetiology involved a focus on hereditary versus acquired deafness. Eugenicists were particularly keen to eliminate the former and to minimize the latter by environmental health management. As we shall see, however, new means of acquiring hearing loss in the First World War shifted the debate into a new medicalized framework.

One particularly aggressive promoter of eugenic ideals was Percival MacLeod Yearsley the Ear Surgeon and London Schools Inspector on otological matters. Since 1893 the education of deaf children had become the responsibility of the taxpayer.[21] As a eugenicist he was committed to pinpointing the extent to which deafness was passed on by intermarriage in order—following Bell's example—to eliminate this costly 'burden' to the London taxpayer.[22] In doing so, Yearsley was methodologically committed to discerning which London school children had non-hereditary forms of deafness both so as to exclude them from his eugenic statistics, and (less centrally for him) to pinpoint which were the diseases and infections that needed most medical attention to minimize future cases of acquisition. His own figures showed that cases of acquired deafness (c. 97.3%) were much more commonplace than those that he claimed to be hereditary (c. 2.7%).[23] As we will see in Chap. 6, however, MacLeod Yearsley devoted vastly more than three per cent of his research time to hereditary deafness.

However, once the First World War began in 1914, rather different forms of acquired deafness among adults became much more of patriotic interest to the medical profession. As Coreen McGuire demonstrates in her Ph.D. thesis, both medical and social attitudes to hearing loss in Britain were significantly changed during the First World War as the deafened, especially the 'war-deafened' emerged as a new category of sympathetic concern. Just a few months into the war in November 1914, The National Bureau for the Encouragement of the Welfare of the Deaf offered its services to the War Office to deal prospectively with 'soldiers and sailors who may suffer from deafness'. Specifically, it considered a major new source of hearing loss would be the way that the 'heavy percussion of modern artillery was likely to affect hearing'. Indeed, this subject came up regularly in the Bureau's minutes during the early years of the war, although the Bureau was not alone in seeking to highlight this concern, and ultimately was not as influential as a newly galvanized medical community.[24]

Strikingly, then, the pre-war pre-occupation with the economic cost of non-standard hearing due to ill-health or inheritance was now set aside to devote energy instead to the new blameless and indeed morally valorised category of the 'War-deafened'. While this group was later defined by self-identification in post-war compensation claims, initially the public identification of this group was a medical interest expressed through publication in clinical journals. For example, in September 1917 Macleod Yearsley noted in the *Journal of Laryngology, Rhinology and Otology*, that over the preceding 3 years many cases 'illustrating the effects of modern high explosives upon the organ of hearing' had been collected from the battlefront in France and Belgium. Cases he had seen ranged from temporary shock-related deafness to a 'permanent injury' to hearing. Yearsley deemed those 'lucky' who had only acquired temporary deafness due to middle-ear conditions; and indeed this was the group for which medical intervention had the strongest prospect of providing assistance. Those rendered permanently deaf were, as ever, beyond the scope of medical care to heal.

However, the title of his piece 'An Air Raid Case' indicates it was not only male soldiers at the battle front who encountered this condition. By June 1917, when the bombing of London by zeppelins had been succeeded by deadly German Gotha bombers, even ordinary civilians otherwise visibly uninjured by falling missiles could find themselves subject to the effects of brutal wartime explosions. In ways that contrast to the general indifference of the medical profession to factory and other workplace explosions in the preceding decades, there was a moral imperative to capture the experiences of the war-deafened as a special category and valorize them with detailed medical attention. Yearsley related the case of Miss X, a teacher in an east London school aged 27, who 'although not injured at the Front, suffered nevertheless in the service of her country.' During the air raid that month she was ushering her class into the school basement when a bomb landed near the closed front door where she was standing: the explosion burst the door open and she was knocked down, injuring her head. Her experiences of resulting hearing problems were recorded by Yearsley in some detail.[25] This new interest is presumably one of many reasons that MacLeod Yearsley's concern with the eugenics of deafness had all but disappeared by the post-war period.

The war-deafened,—during the war at least—unarguably constituted a new general category of deafness to be investigated and cared for. While 'war' was the notional causal factor rather than ill-health or inheritance, this condition covered all kinds of deafness from mild reversible hearing loss to absolute and permanent loss of hearing. *The Lancet* and *The British Medical Journal* among other journals dedicated much interest to cases of such 'war deafness', albeit to publish research wherever no medical amelioration could be accomplished. While the British Government provided some pension provision support for them, it was, as Coreen McGuire shows, charities such as the dedicated Deafened Ex-Service Men's Fund to whom many looked to find ways of coping as War Veterans. When the postwar Industrial Training Scheme was launched in 1919 the National Bureau especially emphasised the virtues of 'deafened workers'. Experience had apparently shown that their 'freedom from the distractions of talk and noise' tended to make them more productive than others.[26]

Yet as the Bureau's successor body noted in 15 years after the war, the overall situation of the 'war-deafened' had not been helped by the advent of widespread telephony and wireless that had 'revolutionised life' for many. Even as long respected subjects of hearing loss, the deafened war veterans faced a world in which even their military heroics gave them no equal footing with the blinded for whom these purely acoustic media had been a great boon.[27]

4 Conclusion

The categorisation of deafness as a form of hearing loss before or during the First World War covered many sorts of experience. While various vocabularies for understanding deafness as hearing loss emerged these were closely entangled with the contexts of social position, and the broader politics of who would represent the concerns of the 'deaf' as an overall group. While the deaf/hard of hearing themselves were concerned with degrees and kinds of deafness and the various means of communicating (hearing or otherwise) open to them, the medical profession was more professionally interested in understanding the causation of deafness and hearing loss: ill-health, inheritance and latterly combat. Certainly, however, there were some key changes in our period. It is significant that the former group dropped 'deaf and dumb' from their lexicon, while the medical fraternity introduced the category of 'war-deafened'. Most obviously not all categories of deafness seen as hearing loss were equally valued: we can see from the stories above, those linked to disease or hereditary factors were less broadly valorised than war-induced forms.

Notes

1. Harriet Martineau, 'Letter to the Deaf', *Tait's Edinburgh Magazine*, April 1834, 174–179.
2. Irene Leigh *A Lens on Deaf Identities* (Oxford & New York: Oxford University Press, 2009).
3. We are grateful to Mike Gulliver for permission to see his piece, '"Deaf-mute vs. Semi-mute": Debating deaf abilities, identities and destinies on the eve of Oralism' that is currently under consideration for publication.
4. For research on the 'tragedy' of blindness, see Julie Anderson *War, Disability and Rehabilitation in Britain: 'Soul of a nation'* (Manchester: Manchester University Press, 2011).
5. Jennifer Esmail, *Reading Victorian Deafness: Signs and Sounds in Victorian Culture* (Athens, Ohio: Ohio University Press, 2013), 9.
6. Jan Branson & Don Miller, *Damned for their Difference: the cultural construction of deaf people as "disabled", a sociological history* (Washington D.C.: Gallaudet University Press, 2003), 145. Esmail, *Reading Victorian Deafness*, 4.
7. Alain Corbin, trans. Martin Thom, *Village Bells: Sound and Meaning in the Nineteenth-Century French Countryside*, (London: Papermac, 1999), 40, 92–93, 95, 194, 288, 307. For other examples of scholarship on sound worlds that take standard hearing capacity for granted but mention hearing loss in passing, see Daniel Morat (editor) *Sounds of Modern History: Auditory Cultures in Nineteenth and Twentieth century Europe* (Oxford & New York: Berghahn, 2014).
8. For a discussion of poetry by the non-hearing deaf, see Esmail, *Reading Victorian Deafness*, 22–68.
9. Emily Varndell, 'The Lament Of The Deaf Lady,' *The Literary Gazette: A Weekly Journal of Literature, Science, and the Fine Arts* no. 1747 (Jul 13, 1850): 478. http://search.proquest.com/docview/5134212?accountid=13651 (last accessed October 22, 2015).
10. Lancaster J. Ball, 'A Deaf Man's Monody In Spring.' *The Wesleyan-Methodist Magazine* (05, 1886): 370. http://search.proquest.com/docview/3041221?accountid=13651 (accessed October 22, 2015).
11. David Low, 'A Matter of Hearing,' *Spectator*, (New Zealand) July 23, 1908, p. 5.

12. 'Deaf As a Post: the raikes' progress *versus* the Post Office pilgrims' progress, *Fun,* July 9, 52 (1890), 15.
13. The periodical itself had many changes of title—later *British Deaf Monthly*, but previously the *Deaf Chronicle, Deaf and Dumb Times.* These changes of title reflected both the diplomatic difficulty of naming in an inclusive fashion and also the financial significance of engaging as large a purchasing readership possible. See discussion in Chap. 6.
14. *The London Reader* published an article in 1878 about an American 'deaf-mute telegrapher' who had become deaf just after he had completed his training, and went on to work in telegraphy very successfully for the rest of his life. 'A Deaf-Mute Telegrapher,' *The London Reader: of literature, art, science and general information*, April 6, 1878, 536.
15. In the evidence to the same Report, we read that of three women employed on one farm, one was a single woman aged 35 'deaf and dumb'. 1868–1869 [4202] [4202-I] Commission on the Employment of Children, Young Persons, and Women in Agriculture (1867). Second report of the commissioners, with appendix part I, 186, 658.
16. George Frankland was later the author of texts such as *The Pure oral method in relation to the environment of the deaf* (1901). See biographical details in British Deaf Mute, 5 (1896), 290-1 and Dominic W Stiles, 'George Frankland, Deaf Journalist (1866–1936)'. http://blogs.ucl.ac.uk/library-rnid/2016/06/10/george-frankland-deaf-journalist-1866-1936-brilliant-scholar-deep-thinker-and-one-of-the-finest-writers-of-prose/.
17. George Frankland, 'By What Name?' *The British Deaf-Mute* Vol. 5, 1895–1896, 11–12.
18. Frankland, 'By What Name?', 11.
19. Frankland, 'By What Name?', 11. As an indication of the mood for unity, this article was the followed immediately by one on the newly-formed National Association of Teachers of the Deaf, with the stated aim of bringing together all those who taught in deaf schools, across 'all systems; the pure oral, the oral, the sign-manual, the combined, the dual' W.S. Bessant, 'The National Association of Teachers of the Deaf,' *The British Deaf-Mute*, 5 (1895–1896) 12–13, 13.
20. Brian H. Greenwald, 'Taking Stock: Alexander Graham Bell and Eugenics,' *The Deaf History Reader,* ed. JohnVickery Van Cleve (Washington D.C.: Gallaudet University Press 2007), 136–152.
21. Peter W. Jackson, *Britain's Deaf Heritage* (Edinburgh: Pentland Press Limited, 1990), 128–129.
22. See Chap. 6 of this volume for further discussion.
23. His figures were 'Infective Diseases' 34.4% and 'infectious Fevers' 72.9%, with measles and scarlet fever by far the most common causes. Macleod

Yearsley 'The Causes Leading To Educational Deafness In Children, With Special Reference To Prevention', *The Lancet* July 27 1912, 228–234.

24. The topic of 'Soldiers & Deafness' is cited on p. 147, 151–153, 162, 170, 174, 190, 193, 203, 304 of *The National Bureau for Promoting the General Welfare of the Deaf and National Institute for the Deaf Minute book*, Volume 1 (1911–1926). Source: Action on Hearing Loss Library.
25. Very soon after the incident 'Miss X' found not only tinnitus but also reported to Yearsley that she was 'markedly deaf', especially in her left ear. While the tinnitus had mostly gone at her next consultation with him a week later, her 'deafness was diminished but very little' and she could not differentiate between lower frequency tones and distant sounds. It is unclear to what extent her hearing was restored, for Yearsley's main concern was to correlate this to two purportedly similar cases of male officers 'from the Front'. Percival Macleod Yearsley, 'An Air Raid Case' *Journal Of Laryngology, Rhinology And Otology* 32 (1917), 18–19.
26. See discussion in Coreen McGuire 'The "Deaf Subscriber" and the shaping of the British Post Office's Amplified Telephones 1911–1939,' unpublished Ph.D. thesis, University of Leeds, 2016.
27. 'The Deafened by Disease' in the *Report of the Executive Committee to the Council*, Year ended March 31st, 1933. Minute book of the National Institute for the Deaf, at the Action on Hearing Loss Library. Thanks to Coreen McGuire for this reference.

CHAPTER 3

Advice for Managing Hearing Loss

Abstract Although surgeons in the early nineteenth century sought to cure hearing loss, this strategy was gradually abandoned as it became clearer that most forms were incurable. This practice soon retreated into support for hearing trumpets. We then investigate how Harriet Martineau argued against trust in medical advice and put forward her own example as a self-confident 'deaf' person managing her life with a hearing trumpet. This is complemented by a survey of the often far from disinterested advice from hearing aid vendors about what kinds of hearing trumpet or alternatives should be purchased.

Keywords Medical advice · Curing vs relieving hearing loss · Deaf authority · Commercial advice

Nineteenth century adults experiencing hearing loss doubtless sought to maintain some of their pre-existing relationships in the hearing world. How did they manage this aspect of their personal circumstances, whether or not successfully able to communicate? From the previous chapter we cannot assume they necessarily followed Harriet Martineau's suggestion of adopting a new unambiguous 'deaf' identify in their interpersonal communications. Even as others began to label them as 'deaf' or patronised them as 'suffering', the (typically gradual) process of losing hearing did not necessarily lead to distinct changes of lifestyle or self-representation, especially for those who lost hearing later in life.

G. Gooday and K. Sayer, *Managing the Experience of Hearing Loss in Britain, 1830–1930*, DOI 10.1057/978-1-137-40686-6_3

Instead they could seek advice on how to mitigate the effects of their loss, whether by medical, technological, or other social means.

Whatever option they chose, there was plentiful advice—albeit not necessarily useful or disinterested—on how to adapt everyday life to decreased hearing capacity. We discuss in this chapter the kinds of medical treatment, advice literature and assistive devices available to those experiencing acquired deafness as a 'problem'. Following the previous chapter, we can assume that these commercial resources were addressed to a decidedly heterogeneous constituency. Equally their makers and authors would likely have had diverse experiences—whether personally or at second hand—of the challenges of hearing loss. Yet even those who wrote like Harriet Martineau from the standpoint of acquired deafness, there was one widespread normative assumption: those with hearing loss were solely responsible for managing the consequences of this alleged 'problem'. The upshot was that they should partake in a consumer culture of private medical advice, buying books, hearing aids or even (later in the nineteenth century) lessons in lip-reading. Thus would the hearing world be profited and unperturbed by any inconvenience of differential hearing capacities.

Indeed, for physicians, vendors and advertisers of hearing aids there was much to be gained *for them* in professional and financial terms from emphasising hearing loss as the deafened persons' 'problem'. Opportunities to offer advice enabled these constituencies to make careers, reputations and perhaps a substantial income out of their partisan efforts to 'solve' a one-sided construal of the communicative problems of hearing loss. Their advice was embedded in this paradigm of selling a resource exclusively relating to their kind of expertise. We should thus be cautious about labelling those who sold clinical services, advisory literature or ameliorative hardware for 'deafness' as necessarily either professionals or authority figures—such statuses were moot and contingent on their practical success.

1 Medical Advice Literature

Throughout the nineteenth century there was plentiful advice in many periodicals, newspapers and books on how 'deaf' persons could sustain a capacity for hearing in the face of many causes of lost hearing. The periodical literature included reviews and advertising for these books, as well as aids and purported 'cures' for deafness. One of these was *Plain Advice for all Classes of Deaf People* (1826) by the aural surgeon William Wright (1773–1860).[1] Not formally trained in any branch of medicine, Wright nevertheless became one London's leading aurists (aural surgeons) in the early nineteenth century. As Jaipreet Virdi-Dhesi has noted,[2] Wright's professional aural practice in Bristol began by public lecturing and teaching apparent deaf-mutes the arts of conversation, an unusual practice for an aural surgeon. Having trained his pupil, Anna Thatcher, to be able to converse at court with Queen Charlotte (1744–1818), consort of King George III, Wright secured his elevated class position as surgeon-aurist-in-ordinary in 1817. His fashionable new London practice soon extended to include the Duke of Wellington, and he was even able to publish three issues of his own (short-lived) journal *The Aurist or Medical Guide for the Deaf* in 1825. The putative authority upon which he wrote *Plain Advice* was thus advertised on the frontispiece was not only as 'Lecturer on the Anatomy and Physiology of the Organ of Hearing', but as editor of a journal, and as 'Surgeon to Her Late Majesty Queen Charlotte'.[3]

With such socially unrivalled credentials, Wright's *Plain Advice for all Classes of Deaf People* was framed both as a treatise for medical 'practitioners' and as a means of prompting 'deaf persons' to seek assistance, implicitly (we must assume) to advertise his own practice. Wright complained that many acquiesced in deafness instead of seeking medical intervention: by such 'obstinacy' many otherwise healthy people had either been 'deprived' of a useful social role, or 'snatched' from the position of family breadwinner.[4] Wright thus laid out some physiological and nervous origin causes of deafness and his alleged 'cures' for everything from the condition of the 'deaf and dumb', through to temporary deafness due to catarrh and wax, and the surgical treatments and pitfalls for specific aural conditions and diseases. His proposed remedies included static electricity and galvanism, although prudently avoiding public detail he emphasised the inadequacy—sometimes fatally disastrous—of rival approaches, and the sufficiency of his own capacities to relieve deafness. Readers were clearly to infer that they were simply not qualified to attempt self-management of their hearing loss.[5]

Furthermore, Wright's claim to authority as an aurist was premised on the sheer numerical scale of his successes. Of the 1500 patients across all classes he had seen, ranging in ages from 20 months through to 85 years, he claimed to have cured 496, relieved 380, and partially relieved 290. Even though he admitted that he was unsure of the outcome for 210 and had been unable to help 124, Wright claimed his practice to be so successful that high ranking medical professionals apparently referred their patients and family members to him. He thus warned readers against the 'lower walks' of the profession who used 'obsolete, or improper method', and often 'injure the organ of hearing' so much, that the case was 'rendered incurable'.[6]

Nevertheless, defending such territory for the specialist aurist claiming to be able to cure or relieve almost all forms of hearing loss was not straightforward. As Jaipreet Virdi-Dhesi has pointed out, Wright spent much effort on policing his professional prerogatives against those he dismissed as 'quacks' in aural surgery, notably John Harrison Curtis.[7] Curtis was an Aurist and Oculist whose writing in the 1830s challenged both the presumption of Wright as a physician and offered an alternative to surgery for the long term management rather than any 'cure' of deafness: the hearing trumpet. In his *Observations on the preservation of hearing and on the choice, use and abuse of hearing-trumpets etc of* 1834.[8] Curtis claimed that there were 60,000 deaf people in London, many of whom had 'never sought any medical advice'. This, like Wright, he attributed to 'sheer negligence', but more characteristically from the belief that 'nothing can be done for them, coupled with a fear of being made worse'.

Without naming his adversary, Curtis implicitly challenged Wright's contention that clinical methods alone were sufficient to manage acquired deafness. Curtis instead joined the commercial forces of the hearing aid vendor. As Virdi-Dhesi points out in her Ph.D. thesis, Curtis acknowledged that for those whose hearing loss could not be reversed by clinical means, the hearing trumpet was the obvious tool to be used, made available to the poor at minimal cost.[9] Curtis had some of his own models to offer his readers, especially a telescopic hearing-trumpet that could readily be compressed and stored discreetly in a 'small case for the pocket':

> ...this trumpet as well as my various instruments, may be had of Mr. S Maw, 11 Aldersgate St, acoustic instrument-maker to the Royal Dispensary for Diseases of the Ear) from its portability, convenience, and cheapness, is now generally preferred to all others....[10]

Conceding that the 'interesting science of acoustics' was too broad to treat systematically, Curtis focused instead on his inventions for the 'assistance of the incurably deaf.' His fertility of invention, and brazen borrowing and self-promotion were both fully displayed in this book. Not only did he market small ear-caps to collect sound from different parts of the room—allegedly 'very serviceable and agreeable' for deaf persons while 'eating, reading or otherwise engaged': he also promoted French artificial ears, typically made of shells and the rather heavier German spring-mounted 'silver ears'. In the fifth edition of 1837 he elaborated upon his 'latest contrivance', the acoustic chair, with its distinctive social and hygienic benefits. For this the seated listener would hear sounds transmitted from the opposite side at which a conversationalist spoke: 'thus avoiding the unpleasant and injurious practice of the speaker coming so close as to render his breath offensive.'[11] Here then was a solution to the awkward bodily proximity that annoyed the discomforted deaf listener when concerned interlocutors spoke too close for comfort or hygiene.[12]

The success of Curtis's practice of selling hearing devices at the Royal Dispensary for Diseases of the Ear is a clear indication that not all trusted to Wright's injunctions to accept only surgical methods. Revealingly, the final work that Wright produced in 1860 conceded that some such mediated hearing devices, especially hearing tubes, were necessary 'to make very deaf persons hear'. Yet he maintained to the end that hearing trumpets were inevitably 'very injurious to the sense of hearing' since the sheer force of sound at the tympanic membrane would be so damaging as to require ever stronger trumpets until all hearing capacity disappeared entirely.[13] Nevertheless Wright had to concede the utility of assistive devices as the high-profile and sometimes deadly failures of aural surgery became apparent by the 1840s.[14] As one eminent Dublin aurist Sir William Wilde, summarised the status of cures for deafness at around the same time: 'There are two kinds of deafness: one is due to wax and is curable, the other is not due to wax and is not curable.'[15]

Such conflicting views among self-styled medical practitioners, as well as their obvious fallibility, inevitably raised challenges about their authority. Deaf subjects frustrated by the often futile and expensive efforts of physicians to 'cure' them, thus had to turn to other sources. If we look beyond medical treatises, indifference or even outright challenges to the injunction to 'cure' deafness are not so hard to find.

2 Advice from a Deaf Authority

One prominent critic of medical prerogatives in handling deafness was the eminent writer and sociologist Harriet Martineau (1802–1876). When she published her oft-cited 'Letter to the Deaf' in 1834,

Martineau's reputation was becoming well established through her first book length writings on political economy and taxation.[16] On smaller scale topics, she was also a habitual writer of letters published in journals and elsewhere. Her 'Letter to the Deaf' appeared in both formats initially appearing in *Tait's Edinburgh Journal*[17] in 1834 and then republished in her widely read essay collection *Miscellanies* two years later.[18]

Martineau's challenge to masculine medical authority is well known in her iconoclastic patient-centred *Life in the Sickroom: Essays by an Invalid* of 1844.[19] What is less often recognised is that a whole decade earlier, Martineau's 'Letter to the Deaf' had openly questioned her physicians' competence and effectiveness in treating the hearing loss that she had begun to experience aged 17. Having endured their repeated unsuccessful attempts to 'cure' this deafness, Martineau instead boldly took on the mantle of expert adviser herself. She advised her readers that the incurably deaf should not to hold out any false hopes that physicians had any means to ameliorate their loss:

> ...if we can find physicians humane enough to tell as the truth: and where it cannot be ascertained, we must not delay making provision for the present...The physician had rather not say, as mine said to me, "I consider yours a bad case."... We sufferers are the persons to put an end to all this delusion and mis-management. Advice must go for nothing with us in a case where nobody is qualified to advise. We must cross-question our physician, and hold him to it till he has told us all.

Readers were enjoined to take matters into their own hands and defy the medical pathologisation of their situation. Martineau advised them to treat their (presumptively partial) deafness unembarrassedly as a new way of life and declare it as their new identity:

> When every body about us gets to treat it as a matter of fact, our daily difficulties are almost gone; and when we have to do with strangers, the simple, cheerful declaration, "I am very deaf," removes almost all trouble.

In the rest of her 'Letter to the Deaf' little mention is made of medical intervention. Martineau recognised that many readers would be 'rather surprised' at her notion that they should be formulating their own 'plans, and methods, and management' for their deafness. She gently chastised those 'too apt to shrink' from regularly taking in hand their 'own case'. But only if they took up their own case could they overcome the vices too often attributed to the 'partially deaf' (as she called them): unscrupulousness about the truth, irritability or moroseness; suspicious, and 'ill-mannered'. By contrast she encouraged the deaf to adopt the highest level of decorum so that they might altruistically be the best company to others: 'We must struggle for whatever may be had, without encroaching on the comfort of others.'

Chief among the many aspects of her self-help manifesto was the necessary acquisition of a hearing trumpet. Martineau professed shock at how 'seldom' a deaf person was seen using one. She had heard many explanations for this: some found that the hearing trumpet made conversation sound 'disagreeable' or entirely of 'no use', perhaps because their friends 'did not like it' being used. While some claimed it did not enable them to hear 'general conversation', others felt it was not needed in a noisy environment, because they heard 'better in a noise'; others again felt that it was not needed in a quiet environment, because 'we hear very fairly in quiet'. There were many other reasons just as unconvincing to her which she believed just to be 'excuses'. As Martineau opined of her own much more successful experiences of the hearing trumpet:

> The sound soon becomes anything but disagreeable; and the relief to the nerves, arising from the use of such a help, is indescribable. None but the totally deaf can fail to find some kind of trumpet that will be of use to them, if they choose to look for it properly, and give it a fair trial.

The burden thus fell (as ever) to the partially hearing person to take responsibility in these matters: they should make the effort to choose a device among the many in the market place and dedicate time to using to it before casting any judgement. Nevertheless, she admitted that the hearing horn did not support conversation so well when the horn amplified background noise as much as the conversation, or in dark conditions 'when the play of the countenance is lost to us'; the latter point was a clear indication that some reliance upon lip-reading was anticipated. But these intermittent difficulties were not reasons to dismiss the hearing horn outright as some had done:

> To reject a tête-à-tête in comfort because the same means will not afford us the pleasure of general conversation, is not very wise...As for the fancy, that our friends do not like it, it is a mistake, and a serious mistake. I can speak confidently of this.

Much has been made by various scholars of Martineau's less than complete adherence to her own professed practice of demure trumpet usage for unobtrusive participation in conversation, often expecting individuals to speak directly into her trumpet.[20] And it is not clear how many deafened readers followed her advice to buy a hearing trumpet when medical advice failed. Nevertheless, after her death Martineau's enthusiastic advice for the hearing trumpet gave hearing aid vendors such as Hawksley a pretext to advertise 'Martineau' type instruments, aiming thereby to profit from the name-recognition in her widely-read advice.[21] Thus we turn next to the cultural representation of hearing loss and the advice contained in the advertising of hearing aids.

3 Commercial Advice on Commodified 'Solutions' to Deafness

Despite Martineau's efforts, Victorian cultural representations of hearing loss mitigated by hearing trumpets were drawn from the usually unsympathetic perspective of hearing culture. Hearing loss was depicted as somehow alien, possibly threatening to social order in ways epitomized by the trumpet itself. Consider, for example, Thomas Hood's 'Tale of a Trumpet', which appeared in the *New Monthly Magazine and Universal Register* in 1841, 7 years after Martineau's Letter. Hood's poem told of a deafened old woman, Dame Eleanor Spearing, who, on acquiring a hearing trumpet, gained a supernatural degree of hearing and her intercepted gossip thereafter caused trouble far and wide. Her fate was to be drowned by angry villagers and her hearing trumpet crushed. Hood's satirical suggestion was that deaf women and assistive hearing technologies were a disruptive combination.[22] While Hood thus captured the hearing world's anxieties about subversive hearing trumpets, the advisory advertising literature for deaf people played up instead the anxieties of hard-of-hearing people in order that they should be worried into purchasing hearing trumpets.

In contrast to Martineau's upbeat assessment, the everyday experiences of hearing loss, not least the sense of social isolation, were taboo subjects rarely discussed publically by those who lived with it.[23] This was epitomized in the culture of advertising for 'cures' for hearing loss. Premised on the norms of hearing culture, such advertising instead played on the customary social anxieties generated by the onset of deafness. Such anxieties might only have been created by such advertising, but that was the point: much advertising was designed to make the hard of hearing (and others) reimagine their hearing loss as *their* burden. It was not an issue to be mitigated by others adapting their conversational behaviour, but rather a problem to be eliminated by deafened people *purchasing* a solution to sort it out themselves. In nineteenth century popular culture 'hearing loss' was most commonly associated with the contexts of old age and 'infirmity'.[24] It was clearly most profitable for vendors to advise the purchase of a hearing device among the wealthy elderly: still mourning for their lost hearing, this constituency was most easily persuaded to purchase a compensatory device.

In 1830, for the price of £1.5.0 one could purchase an 'Acoustic reflector', retained in the user's hand, by an elastic band from 'James Scott M.D'. The loose-leafed flyer used to advertise this device hailed from 369 The Strand London, the same address as James Scott M.D., author of *The Village Doctor; or the art of curing diseases rendered familiar and easy* in 1824.[25] That volume had emphasized the (alleged) curability of certain kinds of deafness by electricity or tobacco smoke, thereby precluding any need for a hearing trumpet. By contrast Scott's 1830 pamphlet showed a distinctive shift away from medical presumptions of 'curing' deafness towards, as for Harrison Curtis, mechanical assistance. To overcome the conventional trumpet's directional limitations, Scott recommended instead his innovation of 'voice conductors' with springs and slides. Although more expensive than the other various trumpets that he advertised, these could be worn permanently on the head hands-free, allow the user to speak to several interlocutors concurrently.[26]

This advice to consider choosing from a multiplicity of hearing assistive devices—as recommended by Martineau—was increasingly common from advertisers later in the nineteenth century. This was not just recognition of the many different kinds and degrees of hearing loss increasingly understood to characterise acquired deafness. It also acknowledged growing expectations of users to exercise discretion about the cost, portability, aesthetics and visibility of their devices.[27] The lattermost issue is very important in light of Martineau's advice that the hard of hearing should announce their deafness by a noticeable hearing trumpet. Countervailing advice for the more self-conscious was readily available from the Hawksley *Catalogue of Otacoustical Instruments to Aid the Deaf*, first produced in 1869, representing hearing loss as an 'affliction' to be healed. This advice about disguised devices comes from the1895 catalogue:

> Sensitive persons, particularly ladies, have an aversion to advertising their affliction in public by the use of many of the usual forms of hearing instruments. To meet this very natural objection, such instruments have been ingeniously combined with fans, parasols, umbrellas, muffs, handbags or reticules, bouquet holders, opera glasses, &c. Other instruments are attached to the head and ears, and may be concealed by the cap, hat, bonnet or hair…

For gentlemen disinclined to communicate their growing deafness, Hawksley supplied walking sticks and umbrellas fitted with 'powerful sound collectors'. His company could even supply similar assistance to be adapted for field glasses or the inside of the ordinary silk hat.[28] The clear indication was that not only could relief from conversational awkwardness be discreetly accomplished in almost every social situation: this needed to be achieved by purchasing more than just one hearing aid. This was evidently a 'solution' only for the most affluent, and a clear indication that a demand was being engineered for discreet invisibility: the complementary approach to conspicuous consumption.

Just as medical practitioners had before them, such high street vendors of hearing aids positioned themselves as advisory authorities on the alleviation of deafness. To manage their customers' expectations about the limited nature of their services, the more respectable companies warned that simply buying a hearing aid was not enough: they also needed to do work to assimilate this device into their ordinary conversation. Consider an advisory pamphlet produced by the John Bell & Croyden company (founded 1903) after suggestion that aids to hearing can be fitted with as much 'efficiency' as glasses. The advertising narrative stressed that aids to hearing 'must of necessity call for certain perseverance on the part of the user to obtain the utmost benefit'.[29] This typically meant that time and patience was required to adapting 'deaf' hearing practice to the new sounds heard through the device and the new constraints it imposed on communicative engagement. While fulfilling a very middle class form of self-improvement, the purchasing of hearing aids also required much more advisory literature than of glasses from an optician since the former were so very much more complex to use.

By contrast, less scrupulous mail order advertisers tended to play down the effort required of the new hearing aid user. Playing up the easy of usage typically emphasizing instead that until they bought a hearing aid they would be an 'irritation' to others. A survey of classified advertisement in the late nineteenth century newspaper reveals the widespread nature of mail order advertisements for diverse assistive equipment such as 'deaf aids', 'deaf instruments', 'speaking tubes (or trumpets)' and 'hearing tubes (or trumpets)', and corrective devices like 'artificial ear drums' and 'sound discs' for insertion into the ear.[30] These press advertisements certainly reflected the contextual specificity of different kinds of hearing assistance needed for sermons in the church/chapel, entertainment in concerts and theatres, and for conversation in the parlour or public assembly, a mix of social aspiration and stress on sociability. But they typically offered what appeared to be very easy, if expensive, solutions.

Sometimes, they offered effortless curative creams/liniments to eradicate deafness (along with a remarkable array of other ailments). Others promised simple alleviation that needed no expert attention from an aurist or personal attention from a high-street vendor. One advertiser in the *Penny Illustrated Paper* for 1911 even claimed that by wearing its 'marvellous discovery' deaf people need no longer go about 'wearing the strained pathetic expression caused by their affliction'—this affliction being a 'positive burden to themselves and those around them.'[31] Without the credentials offered by more reputable high street vendors, mail order advertisers sought to lure as many possible customers into their commercial enterprise, playing down how difficult it might be to get any subsequent benefit from their devices—let alone a money-back guarantee if they brought no benefit. They were quick to blame users, however, if these devices did not seem effective.

4 Conclusion

Notwithstanding conventional clichés, loss of hearing in Victorian life was more culturally fraught than loss of sight. In the advisory literature on deafness discussed above, the moral pressures were enormous. As we shall see in Chap. 6, both trustworthy manufacturers and suppliers, as well as the opportunist vendors often criticized as fraudulent 'quacks'[32] tapped into the dominant scripts of Hearing culture, and created a market for aids other than the ear trumpet, by playing on the sense of isolation that the deafened and hard of hearing might experience unless they purchased the appropriate hearing aid for churches, concerts etc. Yet we should not 'read' the experience of deafness from the self-interested preoccupations of those who sought to sell hearing assistance. The hard of hearing did not necessarily 'suffer' without intervention from aurist or acoustic emporium, nor did they experience uncommunicative 'silence' without a hearing aid. Such advertisements reflected the values of the hearing world in which the deafened were expected to participate and to adapt. Such indeed were the politics of attempting to 'pass' as hearing for those who were 'hard' of hearing in a hearing world.

Short-sighted Victorians might have pondered as much as their deafened counterparts how little reliable medical advice was available to them; the only exception was for loss that was temporary and reversible, e.g. by localised infection, but not as a results of more debilitating ailments. Each might have pondered the various designs of technological assistance they could purchase on the high street—spectacles/eyepieces or hearing aids. While both might have been ambivalent about this technocratic and commodified approach, it was the hard of hearing subjects who experienced the much more stigmatized and emotionally complex position. They had more pressure upon them to mitigate their condition, and the more opportunists offering them not entirely disinterested advice on how to become nearly 'hearing' again. But myopic individuals were not constantly assailed by newspaper advertisements inducing a sense of guilt if they did not seek to become fully-sighted again. Nor did they face exclusion from social conversation if they opted not to purchase spectacles. As we shall see in the next chapter, the presumption was often, indeed, that deaf people were not troubled by short-sightedness, and could see easily enough to lip-read or at least read visual signals from gestures to assist in conversation.

Notes

1. William Wright, Plain *Advice for all Classes of Deaf People, the Deaf and Dumb, and those having Diseases of the Ears* (London: Callow and Wilson, 1826). Bergman, 'William Wright, aurist: nineteenth century pneumatic practitioner and a discoverer of anesthesia', *Annals of Otology, Rhinology and Laryngology*, 103 (1994), 483–486.
2. See blog posts by Jaipreet Virdi. http://jaivirdi.com/2013/01/23/william-wright-miss-hannah-thatcher/#more-1605 (last accessed 14th Dec 2016) and by Dominic Stiles. https://blogs.ucl.ac.uk/library-rnid/2013/02/15/treating-deafness-hannah-thatcher-william-wright-and-the-danger-of-thin-shoes/ (last accessed 14th Dec 2016).

3. William Wright (editor) *The Aurist; or, Medical Guide for the Deaf... with translations, and an analysis of foreign works on the subject* (London, 1825) Volumes 1–3; Methods *of treating deafness, diseases of the ears and the deaf and dumb* (London, 1834); *A Few Minutes' Advice to Deaf Persons; comprising also useful information for the professional world. Being an exposition of the fallacy and inconsistency of the practice of Deleau, Kramer and their imitators in England, or America* (London: James S. Hodson 1839); W. Wright. *Observations upon the application of electricity, galvanism, and electro-magnetism as auxiliaries to medicine & surgery: with a concise account of the diseases, &c. in which they have been, or may be, beneficially employed* (London: James Ridgway, 1848); *Practical observations on deafness, and noises in the head, and their treatment on physiological principles: shewing the injurious and often irremediable consequences of violent applications, as exemplified in the case of Field Marshal His Grace the late Duke of Wellington* (London: John Wesley, 1853); *Deafness, and diseases of the ear: the fallacies of present treatment exposed, and remedies suggested. From the experience of half a century* (London: Thomas Cautley Newby, 1860).
4. Wright, *Plain Advice for all Classes of Deaf People*, 88–89.
5. Wright, *Plain Advice for all Classes of Deaf People*, 86–88, 97–101, 153–154.
6. Wright, *Plain Advice for all Classes of Deaf People*, 119–120.
7. Jaipreet Virdi-Dhesi 'Curtis's Cephaloscope: Deafness and the Making of Surgical Authority in London, 1815–1845,' *Bulletin of the History of Medicine* 87.3 (2013): 349–379. "Not to become a breeding ground for medical experimentation:' Examining the Tensions between Aurists and Educators for the Deaf, 1815–1830,' *British Deaf History Society Journal* 15.4 (2013): 8–13.
8. John Harrison Curtis, *Observations on the preservation of hearing and on the choice, use and abuse of hearing-trumpets etc* (1st ed 1834 5th ed London: Henry Renshaw, 1837), iv.
9. Jaipreet Virdi-Dhesi, 'From The Hands Of Quacks: Aural Surgery, Deafness, And The Making Of A Specialty In 19th Century London' (Unpublished Ph.D. thesis, University of Toronto, 2014), 109–110.
10. John Harrison Curtis, *Observations on the preservation of hearing*, 47.
11. Curtis, *Observations*, 47.
12. For broader discussion of spatial factors in communicating orally, especially the issue of bodily proximity as potentially nurturing and also threatening, see Bryan Lawson, *The Language of Space* (Amsterdam: Architectural Press, 2001).
13. William Wright, *Deafness and Diseases of the Ear*, 242–247.
14. Virdi Dhesi 'Curtis's Cephaloscope'.

15. Quoted in George Cathcart, 'The Alleviation of Chronic Progressive Deafness', *The Lancet* vol. 205, 9 May 1925, 968–972. William Wilde was the father of Oscar Wilde.
16. Harriet Martineau, *Illustrations of taxation* (London: Charles Fox, 1834) 5 volumes; Harriet Martineau, *Illustrations of Political Economy;* 9 volumes; (London: Charles Fox, 1834). Michael Hill and Susan Hoecker-Drysdale, *Harriet Martineau: Theoretical and Methodological Perspectives* (London: Routledge, 2002).
17. Harriet Martineau, 'Letter to the Deaf', *Tait's Edinburgh Magazine* (April 1834), 174–179.
18. 'Letter to the Deaf.' Harriet Martineau, *Miscellanies,* vol. 1. (Boston: Hilliard, Gray, 1836), 248–265.
19. Harriet Martineau, *Life in the Sickroom: Essays by an Invalid,* (London: Edward Moxon, 1844). Alison Winter, 'Harriet Martineau and the Reform of the Invalid in Victorian England', *The Historical Journal*, 38 (1995): 597–616. Michael Hill and Susan Hoecker-Drysdale, *Harriet Martineau: Theoretical and Methodological Perspectives* (London: Routledge, 2002).
20. See discussion in Michael Hill and Susan Hoecker-Drysdale, *Harriet Martineau,* especially references in correspondence of Thomas and Jane Carlyle.
21. The first kind of horn deployed by Martineau was mass produced by Kolbe and was apparently 'quite extensively used in England' James A. Campbell, *Helps to Hear* (Chicago: Duncan Brothers, 1882), 28, 36. The second more open-horn model used by Martineau in later life was sold by Maw & Sons; see Elisabeth Bennion, *Antique hearing devices.* (Vernier Press: London & Brighton, 1994). For the Hawksley strategy of using the names of famous deaf people to name its hearing aids, including Martineau, see Chap. 5.
22. John M. Picker, *Victorian Soundscapes*, (Oxford: Oxford University Press, 2003) Jennifer Esmail, *Reading Victorian Deafness: Signs and Sounds in Victorian Culture* (Athens, Ohio: Ohio University Press, 2013), 176–178.
23. See discussion of 'Out in the Cold' in Chap. 4.
24. For example, the trope of the hearing trumpet as indicator of old age in illustrations such as 'Turning over a New Century' *Punch's Almanack for 1913, Punch* online via Open Library. http://www.archive.org/stream/punchvol144a145lemouoft#page/n13/mode/2up. Accessed 4th April 2011.
25. James Scott, *The Village Doctor or the art of curing diseases rendered familiar and easy; with select receipts,* (London: Knight & Lacey, 1824). The sixth edition of 1837 was explicitly authored by Scott. See, for example

advertisement in the *Athenaeum* 1837, 391, 646 (the address is given in the latter). *Village Doctor*, 2nd edition 1825, 124–127.

26. Advertiser Scott, James, 369, The Strand, London, '*Remarks on the application of the Voice Conductors or Hearing Cornets for deafness, and on the use of the Ear Syringe*' folded sheet, paper 3 p., 22.6 × 18.3 cm (Printer: Eames, John, 7, Tavistock Street, Covent Garden, [London]c. 1830). http://johnjohnson.chadwyck.co.uk/pdf/tmp_1884979411819936373.pdf (last accessed online 4th Nov 2015).
27. Instrument makers who produced them will be discussed in more detail in Chap. 5.
28. T. Hawksley *Catalogue of Otacoustical Instruments to Aid the Deaf*, (London: privately published: 1895 & 1909), 5–6.
29. John Bell & Croyden advertising pamphlet [untitled] (London: publisher unknown, c. 1925), 3, (via John Johnson Collection of Printed Ephemera last accessed online 14th Dec 2016).
30. See Jaipreet Virdi-Dhesi Ph.D. thesis.
31. 'To The Deaf, Simple Home Treatment Revolutionises Method of Treating Deafness'. Sent on Trial' *Penny Illustrated Paper*, April 30, 1910, 571; and June 10, 1911, 770.
32. Advert 'Mr Eliasolomons' Voice Conductors for Deaf Persons', *The Satirist or the Censor of the Times*, Feb 17, 1839, 56.

CHAPTER 4

Communicating with Hearing Loss

Abstract This sets up an experiential comparison between the loss of hearing and the loss of vision to show both how they were inter-related through the reliance of many hard of hearing people on vision to conduct communications. This is explored in relation to both the use of lip-reading, often taught by women, and correspondence networks between deaf people, also characteristically suiting the needs of isolated hard of hearing women.

Keywords Visual loss · Hearing aids · Lip-reading · Correspondence networks

The major challenge of hearing loss was indeed the sustenance of interpersonal communication. Harriet Martineau wrote in her 1834 'Letter to the Deaf' that having avowedly 'suffered' years of deafness as an almost intolerable 'grievance', she 'longed to communicate with my fellow-sufferers.' She wrote in the hope that her experiences would benefit others, especially those who had recently lost hearing, in learning how to maintain their position in hearing society.[1] Several generations later the American celebrity, Helen Keller (1880–1968), addressed the issue of communication from the perspective of a deaf-blind person. It was the loss of the 'most vital stimulus—the sound of the voice' that made deafness for her a 'much worse misfortune' than the loss of sight. Comparing her adult experience to memories of life before her debilitating early

G. Gooday and K. Sayer, *Managing the Experience of Hearing Loss in Britain, 1830–1930*, DOI 10.1057/978-1-137-40686-6_4

childhood illness she recalled how it was hearing rather than vision that 'brings language, sets thought astir', and socially maintained the 'intellectual company of man'.[2] The frequent experiential comparison between these two forms of sensory loss is explored in the first part of this chapter to articulate the particular disadvantages of acquired deafness that were so typically invisible to all but the most attentive in the hearing world.

The question of how the hard of hearing should communicate with others is addressed in the latter two sections of this chapter. One key starting point is that, notwithstanding the recommendations of advisory literature discussed in the previous chapter, no purchased hearing aid could guarantee restoration of the *status quo ante* in interpersonal engagements. While (pre-electronic) devices had many technical challenges, the alternative of speaking very close to a deafened person had major challenges too, not least managing the potential social awkwardness of close personal proximity. While both seeing and listening alike became easier closer to an interlocutor, considerations of personal hygiene, morality, social class, gender, relative age, and numbers of people in a room affected the viability of a spatially immediate technique for communicating.[3]

Hence the alternative approaches of lip-reading and letter-writing discussed below held considerable attractions of simplicity and hygiene for the hard of hearing—albeit at the cost of extra labour and the requirement for particularized social pre-conditions. While letter writing via journals was already very well-established in Harriet Martineau's lifetime as an effective means of establishing discreet social networks, the development of lip-reading for adults as an alternative to using hearing aids was a distinct innovation of the later nineteenth century. In both domains women played a more prominent leading role than in the medical interrogation of deafness discussed in other chapters.

We begin with a comparison between the experiences of hearing loss and sight loss, to set the topic into comparative perspective.

1 Hearing Loss vs Sight Loss

While it was a common trope in Victorian culture that blindness was a greater tragedy than deafness, this was not necessarily the view of those who experienced both.[4] Revealingly, the views of the deaf-blind were not commonly acknowledged in that discussion, but this was not because they lacked means of communication. Finger-spelling techniques using the sense of touch were well established by the late nineteenth century. And for those who long recalled their early capacity for sight and hearing, as was the case for Helen Keller the comparison was very poignant. As she wrote in 1910 to her long-term friend and co-researcher, James Kerr Love in Glasgow, most of her public work had been undertaken for the blind in response to requests from 'workers for the sightless'. But early in a long friendly correspondence with Kerr Love as an aural specialist, she declared herself to be 'just as deaf as ... blind'. From that standpoint the 'problems of deafness' were 'deeper and more complex' if not more important, than those of blindness.[5] Significantly, as we saw above, Keller identified the human voice as the most fundamental part of communication, and lost access to this was the most challenging aspect of her life.

Having lost the capacity for sight and hearing (although not for speaking) Keller early on learned the technique of finger-spelling on the palm of her hand from her dedicated teacher Annie Sullivan. This was not enough for Keller, however: she later came directly to comprehend Sullivan's speech by using her fingers to trace the movements of Sullivan's lips and the vibrations of her throat. By learning to interpret these Keller not only learned a form of indirect hearing (albeit only for one conversationalist at a time), but also learned to use her own voice to exercise some power of speech. That way she could effectively engage in two way conversations in the hearing world. In the 1920s Keller dexterously extended this technique to feeling the vibrations inside a radio set to appreciate broadcasts of classical music too.[6]

Keller thus ameliorated her early loss of hearing in ways that were easily understood given her prioritization of the spoken voice for interpersonal communications. Clearly we should not over-interpret Keller as necessarily representing the position of all deaf-blind people. Yet it is salutary to note her resilience in finding alternative sensory approaches to communication. Specifically, whereas Keller used touch as a surrogate for hearing, many with hearing loss were reliant on their capacity for vision instead, just as users of sign language were too for daily communications. But reliance on vision as an alternative to hearing was itself potentially problematic since it too was as susceptible to corporeal loss as hearing, especially partial loss in later life. Hence in our analysis of past coping strategies for hearing loss, we should not take for the faculty of sight for granted.

The experiential comparison of hearing loss and sight loss raised by Keller's case needs to explore the implications of partial sight loss too. Indeed, there were close parallels in the commodification of prosthetic solutions to certain kinds of sight loss as there were to certain kinds of hearing loss. For those who could afford it, visual assistance had long been possible from glass lenses. Specifically, where visual capacity loss just concerned myopia (short-sightedness) or presbyopia (long-sightedness) there was the relatively simple expedient of donning eye-glasses or spectacles. In his classic study of *Victorian Things* Asa Briggs noted that many an oculist or optician could readily supply suitable bespoke magnifying lenses—whether spectacles, monocle or lunettes—to balance on the nose for the wealthier purchaser seeking to maintain full command of published literature and to observe the niceties of elite socialization.[7]

As Briggs noted, however, the growing the nineteenth century preoccupation with buying glasses for enhanced vision was by no means restricted to the wealthier upper classes. By the last third of the nineteenth century the wearing of spectacles was certainly no longer an elite matter: spectacles were mass-produced just as hearing trumpets had come to be in the wake of Martineau's 1834 'Letter to the Deaf'.[8] There were many cultural reasons for this greater anxiety about 'mending' lost visual capacity which we need not discuss here, but perhaps also concrete professional reasons relating to faster railway travel, reading small print, and sophisticated weaponry in the military services—as well as ever longer factory working cycles. This phenomenon can be captured by the rise of the eponymous 'Snellen' eye-test chart by the Dutch ophthalmologist Herman Snellen c.1861–1862 that rapidly spread across the industrial world. The rapid spread of this test indicates that the rising profession of ophthalmology was anticipating a more industrial scale of operation than the bespoke fitting of spectacles for individuals.[9]

As evidence of the rise of mass spectacle wearing Briggs cites 'Eyes and Eye-Glasses: A Friendly Treatise' by the poet and journalist Richard Hengist Horne, published in Thomas Carlyle's *Fraser's Magazine* in December 1876. Horne wrote to promote the easy availability of spectacles and advised his readers to do business only with the 'educated, practical optician' irrespective of his celebrity or place of abode'. These ordinary high-street 'adepts' would likely do the 'very best that can probably be done' selling a pair of steel-framed spectacles for two shillings which an elite oculist might supply for ten times more. Indeed, Horne argued that the great majority of the population troubled by 'a near-sight, aged or far—sight, and imperfect sight' could be handled by reasonably inexpensive 'eye-glasses'.[10]

They were correlatively warned against both fraudulent hawkers and 'charlatan cheap shops', where glasses bought for a shilling might 'damage your eyes for life', and also against the expense of being 'dazzled and deluded by great names' with royal arms, or 'fine, scientific-looking instruments and lenses' displayed behind plate-glass windows. Since inexperienced purchasers might risk excessive charges by opportunist opticians, he advised readers to avoid those who used the grandiose language of 'learned nomenclatures, science, and technicalities'. Another point of suspicion was any eager oculist's suggestion that they detected 'the squint, the odd or unequal eyes, and the misrepresenting vision, whether of colours or forms': such diagnoses could lead to inappropriate and expensive purchases.

By whatever stratagems these spectacles were sold, Briggs emphasises that such was the growth of the spectacle-making industry in the 1870s that Ophthalmology was becoming established internationally as a new technical sub-discipline linked to medicine—just as the newly emerging discipline of otology was becoming established in studying ear disease. Following German and American precedents, the UK's Ophthalmological Society was set up in 1880 to bring together the leading ophthalmic surgeons; the *Ophthalmic Review* began in 1882, and from 1898 the British Optical Association organised examinations for opticians to qualify as members. However, this organisation was powerless to prevent at least one major Northern town allowing non—members to sell pairs of spectacles at a penny each from a market stall.[11] Access to technologically enhanced vision was thus as potentially fraught with difficult choices about quality and professional ethics as for hearing trumpets and speaking tubes.

Returning to our earlier theme from discussion of Helen Keller above, there was a correlated development in the later nineteenth century that took for granted effective visual capacity as regulated by ophthalmologists. This was the rise of adult lip-reading. As one leading teacher of the art noted in 1902, a high degree of visual acuity was presumed:

> Lip-reading of course depends on the eyesight, and in a general way, if this is good and the light fair, the reader can see the articulation, the pauses, and the entire facial expression so much better if at some little distance from the speaker.[12]

Magnifying spectacles would thus have to be used by any deafened persons if lip-reading were undertaken with eyesight problems. But as we shall see in the next section, more than this was required if lip-reading were to serve hard-of-hearing people as a mode of communication. Much else hinged on both the lip-reader's proximity to their interlocutor and on the quality of illumination that fell on the lip movements to be read.

2 Lip-Reading and its Pitfalls

Lip-reading was discussed occasionally in British deaf journals, as readers and journalists debated the merits of various approaches to communication. Occasionally the use of lip-reading by deaf celebrities was noted to highlight how they had maintained dialogue in the speaking-hearing world. For example, the *Deaf Chronicle* in April 1892 reported of one famed New Jersey native who was a highly effective lip-reader:

> I wonder how many of my readers know that Thomas Alva Edison, the greatest inventor of the world, is absolutely deaf? … Edison has frequently surprised many of his visitors by freely answering all questions put to him. This man of inventions has taught himself lip-reading, and is perfect, seldom making mistakes.[13]

Yet the art of interpreting the visible movements of a speaker's lips was never consensually accepted as an effective or appropriate method of communication for deaf people. As Branson and Miller have noted, the origins of lip-reading were very far from this celebrity milieu in the contested world of deaf education. In contrast to the origins of sign language due to L'Epee in France, the oral system had apparently been developed by Samuel Heinicke (1723–1790) in the Germanic states solely for the early instruction of 'Deaf-mutes', especially children.[14] It is not therefore self-evident that lip-reading would become a mainstream communicative option for partially deaf adults.

One major source of Germanic oralism in the UK was Susannah Hull, born of wealthy medical parents, who opened a private school for the deaf in Kensington in 1863. Initially aiming to teach mixed finger and oral methods, her encounter with Melville Bell's oral technique in 1868 and then his son Alexander Graham Bell in 1872 persuaded her to adopt lip-reading for her teaching and teacher training.[15] Hull was a founder of the Ealing Society for Training Teachers of the Deaf in 1878, which had a primarily female clientele of trainees in lip-reading. Another source of the 'German method' was the Dutch-Jewish teacher William Van Praagh (1845–1907) who introduced it in 1866 as manager of the Jews' Deaf and Dumb Home in London. That appointment in turn led to the Baroness Mayer de Rothschild's establishment in 1870 of the secular 'Association for the Oral Instruction of the Deaf and Dumb' directed by Van Praagh. Following his 1871 plan of Day-Schools for the Deaf and Dumb', the Association appointed him to set up the 'Normal School and Training College' in Fitzroy Square, London a year later. This institution promoted lip-reading as an exclusive 'pure' method. For the next 30 years at least, the majority of British teachers in the lip-reading system were apparently trained there.[16]

Well before the notorious oralist conclusions of the Milan Congress of 1880, there were already substantial projects for education of British deaf children via lip-reading. And even before that, critics railed against the unduly demanding nature of this art. William Benjamin Carpenter, Fullerian Professor of Physiology at the Royal Institution addressed this topic in his *Principles of mental physiology* (1875) recording evidence of the remarkable visual acuity required. Carpenter had been struck by how few 'Deaf-and-Dumb' individuals had acquired advanced lip-reading skills in oralist programmes because only a few had the advanced visual capacities required in lip-reading. That it should have been 'even exceptionally acquired' by one specific group of deaf individuals was for him evidence only of the 'extraordinary improvability of the Perceptive faculty'. It was certainly not evidence of the general utility of lip-reading. Carpenter did not, of course, consider the contemporary evidence of lip-reading skills among contemporary working peoples in noisy factories.[17]

Given such sharp critique even before the Milan Congress, how was training in lip-reading extended from children to partially deaf adults? One answer is that a new generation of instructors emerged that sought a regular income from a deafened adult clientele.[18] Consider the case of Eliza Frances Boultbee (1860–1925) who initially learned lip-reading to educate her 'deaf and dumb' younger sister Anne (1865–1887). After Anne's death, Eliza taught the skill to a broader group of children and adults, and it was c.1898 that we know she apparently became acquainted with Van Praagh's writings.[19] By the time she published *Practical Lip-Reading For The Use Of The Deaf* in 1902, she claimed to have such a broad range of clients that she was obliged teach all day every week day.[20]

While acknowledging the controversial origins of lip-reading in Oralist teaching of children, Boultbee stressed that her main audience was the 'partially deaf' adult. She stressed that while lip-reading could be learned and retained at any stage of hearing loss, ear-trumpet users might need to purchase a series of ever more powerful devices as their deafness increased, until ultimately they would only be able to lip-read anyway. The sheer number of satisfied clients who bought into Boultbee's methods vindicated this approach. One such, previously (we assume) in a dedicated care institution, apparently wrote to her:

> Everyone is delighted. You cannot think what a difference the Lip-reading has made to me. I have not used my ear-trumpet since I came home; formerly, I never thought of going down to a meal without it. Many thanks for the great help you have given me.[21]

Practical Lip-Reading was also written to show family and friends how they shared responsibility for effective communication by facilitating easy reading of their lip movements. Lip-readers might require additional time to process the meaning of visually—processed speech: impatient repetition was distracting and 'annoying to a deaf person'. Nor was shouting at deaf friends helpful: this actually made understanding harder too.[22] Even so Boultbee emphasised that lip readers had to manage their situation with some prudence. *Pace* Martineau, she advised deaf lip readers not to advertise their situation, especially with inexperienced or insensitive conversationalists who would tend to make counterproductive 'hideous facial contortions' in blundering efforts to have their lip movements better read.[23] And since facial visibility was a crucial consideration, she advised lip-readers to choose venues with the best illumination:

> It is so much easier when a good light is on the countenance; not necessarily a full light, or that the face should be directly turned to you. In some cases it will often be found easier to read speech from the profile view, at almost any angle, rather than from a full view of the lips.

Nevertheless, she advised that lip-readers needed to adapt to the great variety of speaking styles and be prepared to laugh at their own mistakes. When sight-reading fatigue set in, she advised that they should take a book, paper or knitting as a diversion, but keep eye contact for when conversation restarted.[24] To protect her professional territory, Boultbee never acceded to Carpenter's suggestion that lip-reading was only feasible for an exceptional few. On the contrary, her apparent continued success in teaching won ever more to lip-reading—and also the approval of Sir James Goodhart, Consulting Physician To Guy's Hospital. It was he as a clinical generalist, rather than any otologist, who wrote the approving preface to her second book: *Help for the Deaf: What Lip-Reading Is* in 1913.[25]

Although this book was generally well-reviewed, reviewers understandably drew attention both to the limits both of lip-reading and to which lip-reading could be learned from a book anyway. Lip reading could only be used for one person at a time, in well-lit conditions, and for people near to hand. These were not limitations to which lip-reading instructors drew attention, especially as these were issues which were not troublesome for users of hearing trumpets.[26]

Indeed another topic that became intractable for deaf people from the 1880s not mentioned at all by Boultbee or other sign language teachers is the telephone. As discussed by Esmail, the telephone raised major problems for many kinds of deaf people: there was no possibility of lip-reading, or sign communication, with this first entirely non-visual communications device. Thus when the first mention was raised in the deaf media about the possibility of a telephone for the partially deaf, there was understandable interest. In 1908 the *British Deaf Times* reported on a 'Telephone for the Deaf' supposedly to be made by the Stolz Electrophone Company of Chicago in a new London factory. The house writer evidently thought this was a 'patent pocket telephone for the deaf' might benefit the partially deaf, although could not benefit the 'totally deaf' who must instead 'adhere to lip-reading if they feel able to stake their faith in it'.[27] While the company did eventually produce an 'electrophone' this was not in fact a telephone for the deaf, but simply a pocket-sized hearing aid with telephonic microphones that was indeed of no use to anyone heavily reliant upon lip-reading.[28]

Lip-reading, as noted above, was not to the liking of all who tried it. So let us consider another complementary strategy for the hard of hearing to communicate.

3 Literary and Epistolary Strategies

Experiments in lip-reading were intractably difficult or fatiguing for some so various writing and reading practices served as a practicable alternative. Such practices ranged from soliciting short written notes to letter correspondence between isolated hard of hearing people, all the way up to professional publishing as journalists. The first of these was evidently familiar for those deaf people with comfortably literate fellow conversationalists and writing resources ready to hand. The additional use of pen/pencil and paper as an alternative or a supplement to oral communication at ordinary social gatherings, for example, was established enough to add verisimilitude to Victorian fiction on life for the deaf.

One example also highlighted, however, the challenges to social etiquette given the greater gravitas customary for the written word. Consider the case of 'The Deaf Colonel', a short story published in *Belgravia: A London Magazine*, (1897) an illustrated quarterly founded by sensation novelist Mary Braddon. The particular deafness of the titular protagonist lead to his friends running 'the risk of being entirely misunderstood' when shouting at him. So he always carried in his pocket a pencil and a roll of paper. A conversation is set out as follows as he addressed a new female acquaintance straining in verbal methods:

> "You would prefer writing, would you not?" he said, deferentially, "It is less tiring," and he drew from his pocket a roll of paper and a pencil.

This was quite a challenge to her, as the author A.H. Begbie, conveyed in her inner narrative voice. When receiving a pencil and paper in one's hand, 'it appears that only weighty thoughts can be worthy of such servants'. She thus nervously turned the pencil in her fingers, almost praying that the ground might swallow her up. She eventually overcame her embarrassment, however, and wrote 'Do you know Florence very well?' Their long conversation about travel and Eastern literature was characterised asymmetrically by his speaking and her response in writing, the latter with increasing fluency and confidence, with their relationship flourishing in consequence.[29] A similar experience about living with lost hearing is apparent in a series of letters to the *Spectator* from 1917–1919. One life-long deaf person, A.C. Dave recalled how awkward was the 'impatience' often shown when he requested a conversationalist to 'use a pencil in reply to a necessary question'. The experience of requesting written substitutes for conversation was evidently commonplace but did not generally make for a comfortable communication.[30] Indeed Dave points out that while the hard of hearing expected hearing people to resort to pencil and paper, the courtesy was not often reciprocated.[31]

For those who found it painfully difficult to engage in either lip-reading or recourse to the scribbled note, a further recourse remained: the letter. Harriet Martineau's Letter to the Deaf in 1834 illustrated the emotional efficacy of the open letter to share widely the consequences of hearing loss.[32] It was thus possible with the help of well-placed and sympathetic friends for some deafened people to pursue their careers mostly or entirely through correspondence. In the world of telecommunications, for example, important partially deaf figures such as Oliver Heaviside and Alexander Muirhead conducted their discussions with colleagues by letter, and hardly ever appeared in public. There is no evidence of them ever trying out a hearing aid or appearing in public. They were instead supported by family members, who typically represented them at public events.[33]

Occasionally through letters to the mainstream press, as well as deaf periodicals we can glimpse the experience of those who declined both hearing trumpet and lip-reading. This can be seen in correspondence *The Woman At Home*, a monthly volume edited from 1893 by Annie S. Swan (hence its subtitle *Annie Swan's Magazine*).[34] In the regular Section header: 'Life and Work at Home', apparently edited by Jane T. Stoddart, women with various concerns wrote in, often pseudonymously, to receive or offer advice or to solicit contact with others sharing the same interests. Early in its existence we see a letter published from 'A lady suffering from that sad deprivation of complete loss of hearing'. This was essentially soliciting customers for her correspondence course in lip-reading: the address given was that of Miss PORTER, Normandy Villas, Chapel Allerton in Leeds.[35] In subsequent discussions, she is not mentioned again, perhaps owing to doubts about teaching lip-reading remotely without direct face-to-face contact. Indeed a few months later the columnist enquired on behalf of another correspondent on hearing loss on the Sussex Seaboard, 'F.L.', whether it was feasible to learn lip-reading by book.[36] The answer appears to have been that it was not.

For another correspondent writing as 'DEPRESSED ONE' the recommendation about lip-reading, was instead to visit Miss Elizabeth Boultbee recognised, as an authority on the topic, even if not correctly spelled:

> I am very happy to be able to give some satisfactory information about lip reading for the deaf and partially deaf. If she would call at Members' Mansions, Victoria Street, S.W. and ask for Miss Boltbee [sic], she will receive all information about this interesting science. I am sure that Miss Boltbee would be able to do her a great deal of good.[37]

Yet the unsatisfactory, or at least underexplored virtues of lip reading, became evident in later issues, when a correspondent writing in as ‘Out in the Cold’ declares that hearing loss has left her feeling very much an outsider. A correspondent fashioning herself as ‘Mag’ soon wrote into assist, suggesting that a hearing assistive device might be the answer.

> [Mag] recommends a very convenient appliance, which has been invented for the use of the deaf, and of which she has some personal knowledge, as it has been used successfully by a member of her own family. If OUT IN THE COLD cares to write to Messrs. F.C.Rein & Son, 108, Strand, London, she can obtain further particulars.[38]

More revealingly, another correspondent fashioning herself ‘Antigone’ wrote in sympathetically to address ‘Out In The Cold’ as one who had also experienced loss of hearing, and found personal correspondence much more rewarding than lip-reading:

> I can truly sympathise with her, being deaf myself, and know well what it is to feel out in the cold. I see in this month’s issue that lip-reading is suggested as an alleviation of the loneliness which deafness necessarily entails. It is a great help to those who are able to acquire it, but it is not every one who can do so; and to sensitive natures it is rather an ordeal to keep the eyes constantly fixed on a person, especially if the speaker be a stranger.

The answer for ‘Antigone’ was not lip-reading but sociable correspondence:

> If I may speak from personal experience, I should say that the greatest, if not the only, alleviation of deafness is to find others similarly afflicted, and to endeavour to brighten their lives. ‘Those who bring sunshine to the lives of others cannot keep it from their own.’ It has been my happiness and privilege to brighten the lives of two other deaf girls, and I can certainly say I have received more than I have given. If OUT IN THE COLD would allow me to write to her, I should be so glad to put her in the way of helping others similarly afflicted.[39]

Warmed to the response received from other female correspondents, ‘Out In The Cold’ wrote in a second time, evidently asking about whether a cure for deafness could be identified. To this the columnist responded by casting doubt on the claimed 'cures' for deafness to be found in contemporary newspaper advertising:

> I am pleased to have a second letter from OUT IN THE COLD. I sympathise very much with her in her affliction. It is quite true, as she says, that few realize what a deprivation is the loss of hearing. I am sorry I do not know of anything likely to effect a cure in her case, since it would appear she has had the best advice ... Very little reliance can be placed on the personal paragraphs which appear in the majority of newspapers.[40]

This approach to networking in fellow women with hearing loss who sought to communicate via correspondence bringing together, inspired further interest from another correspondent:

> Would the young lady who sympathised with OUT IN THE COLD correspond with another reader afflicted with deafness, and in this way shut out from all social intercourse. The address is M.Wooly, Albert Gate, Evesham Worcestershire.[41]

And thus were communication networks established for isolated deafened women scattered across the United Kingdom through the conjoint media of periodical correspondence column and letter writing.

4 Conclusion

This chapter has explored how written communication and lip-reading were important if not straightforward elements within the mosaic of trying to maintain social interaction. Both had their limits: the use of lip-reading required congenial circumstances and fellow conversationalists; while literary strategies implied a withdrawal from conventional sociability. Nevertheless, they suited some among the hard of hearing who preferred not to use—or could not afford to acquire—the hardware of technological assistance. The latter topic will be discussed in the next chapter, the point being explored further that, for those experiencing hearing loss, it was not—contrary to the rhetoric of vendors—*singularly necessary* to obtain a hearing aid. For many a continuation of 'normal' life could be maintained by writing and lip-reading, and the latter itself later became a requirement for many in the effective use of hearing aids.

Notes

1. Harriet Martineau's 'Letter to the Deaf', published in *Tait's Edinburgh Magazine*, 1834, and *Miscellanies* vol. 1, 1836.
2. Helen Keller to James Kerr Love, March 31, 1910, reproduced as an appendix to James Kerr Love, 'The Deaf Child from the View-Point of the Physician and Teacher', *The Laryngoscope*, 20 (1910) 596–611, 611.
3. Karen Sayer (ed.) *Victorian Space(s)—Working Papers in Victorian Studies* (Horsforth: Leeds Centre for Victorian Studies, 2006).
4. See discussions in Lennard J. Davis, *The Disability Studies Reader* (London, Routledge 2006). But for a contrasting historical view see 'Which is the higher misfortune—to be blind or deaf?' *Mirror of Literature, Amusement and Instruction* (August 15, 1840), 106.
5. Keller to Kerr Love, March 31, 1910.
6. Helen Keller, with Anne Sullivan and John A. Macy *The Story of My Life* (New York, NY: Doubleday, Page & Co, 1903). Anne Sullivan explains how Helen Keller learned to speak http://braillebug.afb.org/hkmediaviewer.asp?frameid=29 (last accessed 19th Dec 2016) Helen Keller's letter to the *New York Times* on tactile 'hearing' on the radio the New York Symphony Orchestra play Beethoven's Ninth Symphony, in March 1924. http://www.afb.org/blog/afb-blog/helen-keller-letter-on-beethoven%e2%80%99s-ninth-symphony-goes-viral/12 (last accessed 19th Dec 2016).
7. Asa Briggs, *Victorian Things*, rev. ed. (Sutton: Phoenix Mill, Thrupp, 2003), pp. 84–97.
8. See Gemma Almond's Ph.D. research blog 'What did the Victorians make of spectacles? http://blog.wellcomelibrary.org/2016/08/what-did-the-victorians-make-of-spectacles/ (accessed 1st December 2016).
9. Robert A. Scott, *The Making of Blind Men: A Study of Adult Socialization* (New Brunswick (USA) and London (UK): Transaction Publishers, 1991), 40.
10. Richard Hengist Horne, 'Eyes and Eye-Glasses: A Friendly Treatise', *Fraser's Magazine*, December 9 (1876), 698–722, esp. 721–722. For a literary-historical study see Katherine Inglis, 'Ophthalmoscopy in Charlotte Brontë's Villette,' *Journal of Victorian Culture,* 2010, vol 15, no. 3, 348–369. That work cites J.T. *Hudson, Useful Remarks upon Spectacles, Lenses, and Opera-Glasses; With Hints to Spectacle Wearers and others; Being an Epitome of Practical and Useful Knowledge Upon this Popular and Important Subject* (London: Joseph Thomas, 1840). For a more recent historical overview of glass material culture see Isobel Armstrong, *Victorian glassworlds: glass culture and the imagination 1830–1880* (Oxford & New York: Oxford University Press, 2008).

11. Briggs, 110–111. On the somewhat later formation in the UK of an otological counterpart, see 'The Otological Society of the United Kingdom,' *British Medical Journal*, July 8 (1899), 93.
12. Elizabeth F. Boultbee, *Practical Lip-Reading For The Use Of The Deaf* (London: L.U. Gill; New York: C. Scribner's,1902).
13. '[A Famous Lip-Reader]', *Deaf Chronicle*, April 1892, 76.
14. Jan Branson, Don Miller, *Damned for their difference: the cultural construction of deaf people as 'disabled': a sociological history.* (Washington D.C.: Gallaudet University Press, 2003).
15. '*Miss Hull's Life Work', British Deaf Monthly, 1899, Vol. 8, no.90, 113* and Boultbee, *Practical Lip-Reading*, 14–15.
16. Joseph Jacobs & Isidore Harris 'William Van Praagh', *Jewish Encyclopedia*, 1906 Funk & Wagnalls Company, 1907, 401; M. *G, McLoughlin, A history of the education of the deaf in England. Liverpool, 1987, and Dominic Stiles' blog:* http://blogs.ucl.ac.uk/library-rnid/tag/van-praagh/ (last accessed 19th Dec 2016).
17. William Benjamin Carpenter, *Principles of mental physiology*, ed, (London: Kegan Paul 1875, 204). For investigation of working class cultures of lip-reading see discussion in Karin Bijsterveld, *Mechanical Sound: Technology, Culture, and Public Problems of Noise in the Twentieth Century* (Cambridge, MA: MIT Press, 2008). We are grateful to Karen Greenlees for sharing her research with us on this topic.
18. William van Praagh, *Conversational lessons for instruction of deaf and dumb children in lip-reading, speaking, reading and writing*; (London: Wertheimer. Lea & Co, 1881); *Lessons for the instruction of deaf and dumb children in speaking, lip-reading, reading, and writing;* (London: Trübner; 1884); 'Lip-reading' *Quarterly Review for Deaf mute Education*. no. I, London, (1888).
19. We know at least that in 1898 Eliza borrowed a copy of didactic paper Van Praagh's article 'Lip-reading' (1888) from Library of the National Association of Teachers for the Deaf. http://www.boultbee.freeserve.co.uk/bfh/ix.htm See Dominic Stiles' blog on her use of the Arnold library, later transferred to the NID. http://blogs.ucl.ac.uk/library-rnid/2014/09/12/urging-that-the-library-should-be-transferred-to-manchester-the-arnold-library-book-loans-register-1899-1922/ (last accessed 19th Dec 2016).
20. Van Praagh, 'Lip-reading'.
21. Boultbee, *Practical Lip-Reading*, 54, 105.
22. Boultbee, *Practical Lip-Reading*, 89.
23. Boultbee, *Practical Lip-Reading*, 90.
24. Boultbee, *Practical Lip-Reading*, 90–91.
25. E. F. Boultbee, *Help for the Deaf: What Lip-Reading Is* (London: Hodder and Stoughton, 1913).

26. 'Some books of the week', *Spectator*, November 29, 1913, 28.
27. 'Telephone for the Deaf'. *British Deaf Times* 5 (1908), 205. Mara Mills, 'Deafening: Noise and the Engineering of Communication in the Telephone System' *Grey Room* 43 (Spring 2011): 118–143; 'Hearing Aids and the History of Electronics Miniaturization'. *IEEE Annals of the History of Computing* 33, 2 (April–June 2011): 24–45. 'When Mobile Communication Technologies Were New'. *Endeavour* 33 (December 2009): 140–146. See Coreen McGuire's Ph.D. thesis for further discussion of this for the WW1 period onwards.
28. After a major Christmas marketing campaign in late 1912, the Stolz company went into liquidation in 1914. See the Advertisement 'Stolz electrophone' *The Times* Nov 26, 1912, 5. On the liquidation see 'The Stolz Electrophone Company (1913) (Limited)', *The Times*, Mar 25, 1914, 24; Law Notices, March 30, *The Times*, Monday, Mar 30, 1914, 3; 'Misrepresentations In A Prospectus,' *The Times*, Saturday, May 08, 1915, 3. A new completely British company was apparently relaunched in late 1915 'The Stolz Electrophone Co' *The Times*, Nov 09, 1915, 3 with a name change to Stols in 1916 [Advertisement]. *The Times* (London, England), Wednesday, Jul 05, 1916, 3.
29. A.H. Begbie, 'The Deaf Colonel,' *Belgravia: A London Magazine*, May 1897, 78–105, 79, 81–83; ultimately, the eponymous Colonel goes to Paris to 'take the cure' (unspecified) to restore hearing and seeks her hand in marriage. There is no obvious evidence that Begbie had direct personal experiences of deafness. Braddon edited this publication 1866–1867 although after Chatto and Windus took over its circulation began to decline. Laurel Brake and Marysa Demoor. *Dictionary of Nineteenth-century Journalism in Great Britain and Ireland* (Gent: Academia Press, 2009) 44–45.
30. Another is example in literature is Wilkie Collins, *The Guilty River*, (1886) e:text available via Open Library. https://archive.org/stream/guiltyriveranov00collgoog (last accessed 10th Sept 2016) e.g. in Chap. 3, the hearing protagonist is approached by a deaf man—Collins' has the protagonist explain that he can't use the 'finger alphabet' and the deaf man says that he can't either as he's only been deaf a year, so he will ask the protagonist to write his thoughts.
31. A.M. Dave, Letter to the Editor, 'Deafness', *Spectator*, December 29, 1917, 12. 'Personally, after nearly forty years of silence … I have found that the partially deaf are among the most unwilling to speak to these who are worse off than themselves in being unable to hear at all. They will not readily use the hand alphabet or pencil even for the sake of others.'

32. Indeed, Martineau certainly carried on some major relationships strictly by correspondence alone, for example with Florence Nightingale whom she never met in person. See Lynn McDonald, 'The Florence Nightingale-Harriet Martineau Collaboration' in Michael Hill and Susan Hoecker-Drysdale, *Harriet Martineau: Theoretical and Methodological Perspectives* (London: Routledge, 2002) 153–168.
33. Paul Nahin, *Oliver Heaviside: The Life, Work, and Times of an Electrical Genius of the Victorian Age* (Baltimore: Johns Hopkins, 2002) and Elizabeth Muirhead *Alexander Muirhead*, (Oxford: privately published, 1924).
34. *The Woman at Home: Annie Swan's Magazine* Published in London, Jan 01, 1894–Jan 01, 1900. Regular Section header: 'Life and Work at Home' Notice to correspondents: All communications to the department 'Life and work at home' except those intended for 'Over the Teacups' must be addressed to Miss Jane T. Stoddart, Office of THE WOMAN AT HOME, 27 Paternoster Row, London, E.C.
35. *The Woman at Home:* Annie Swan's Magazine Published in London, Jan 01, 1894–Jan 01, 1900, 79. Normandy Villas no longer exists, but was a genuine address in Nineteenth century Leeds.
36. *The Woman at Home:* 331.
37. *The Woman at Home:* 610.
38. *The Woman at Home:* 892. The Rein company's devices will appear again in a later chapter as a somewhat expensive if elegant partial solution to mild hearing loss.
39. *The Woman at Home*: 892.
40. *The Woman at Home*: 517.
41. *The Woman at Home*: 1174.

CHAPTER 5

Selling and Using Hearing Aids

Abstract Hard of hearing people generally wanted to maintain conversations with the hearing without the awkwardness of reliance on lip-reading or writing. We discuss the many kinds of hearing aid available, disguised or elegantly designed for display, suited variously for many kinds and degrees of hearing loss. Two especially eminent London-based hearing aid companies are discussed: the genteel bespoke personal service of the Rein Company and the more medically-oriented mail order service of Hawskley. The more opportunist hearing aid vendors who sought just to profit from hearing loss are discussed through the critical eyes of campaigning journalists who were themselves hard of hearing.

Keywords Rein company · Hawksley company · Advertising vendors

Hard of hearing people that wanted to maintain conversations with the hearing without reliance on lip-reading or writing had one obvious alternative. For those who could afford it, there was the use of a hearing aid or 'aid to the deaf'. As ubiquitous as spectacles for short sight, and often visually more striking, the hearing aid has nevertheless barely appeared in mainstream social histories. This is perhaps because hearing trumpets were akin to clothing, and similarly carried close to—almost as part of—the body as entirely conventional accoutrements. Although Harriet Martineau promoted the hearing trumpet as such an obvious solution

G. Gooday and K. Sayer, *Managing the Experience of Hearing Loss in Britain, 1830–1930*, DOI 10.1057/978-1-137-40686-6_5

for hearing loss, reading her 'Letter to the Deaf' would not reveal much about the skill levels required to converse with it. After all, such assistive devices by no means restored hearing to the *status quo ante*. Whereas lip-reading required visual skills, and letter writing presumed literary skills, the use of hearing aids required learning to 'hear' again in new ways. This new augmented hearing could only be mediated effectively through a prosthesis if carefully chosen to match the specific kind and degree of hearing loss that a deafened person experienced. The extent to which Victorian hearing aid vendors drew attention to these complexities of hearing aid is a point that we explore below.

While lip-reading and correspondence networks were comparatively low profile in Victorian society, the same could not necessarily be said of all hearing aids. Certainly some varieties were disguised (against Martineau's advice) in hair combs, bonnets, top hats and walking sticks to hide hearing loss; others were designed to be inconspicuous accessories e.g. black lace-covered hearing horns for ladies in mourning. But many hearing assistive devices were designed to be of a size and decoration to be distinctly conspicuous.[1] Elisabeth Bennion and Kenneth Berger for example discuss London Domes that were large and ornately engraved; these were clearly designed to appeal to those comfortable with displaying their 'hard of hearing' status to all and sundry. Indeed as mentioned previously, such a visible partial deafness enabled the wearer to follow Harriet Martineau's example of signalling to prospective conversationalists that they should adapt their communicative expectations accordingly, potentially saving time and embarrassment. Our survey of the Rein and Hawksley companies' products explores devices of both kinds.[2]

What Bennion and Kenneth Berger, as well as numerous museum curators[3] have noted of numerous hearing aids in their collections, was an enormous variety of design and operation. Their manifold aids were shaped to amplify sound to varying degrees for different kinds and experiences of deafness, and for use by different degrees of wealth, manufactured in a range of materials and deployed in various social and cultural contexts.[4] Many aids were quite straightforward, fixed, simple trumpets in gunmetal, of the sort recommended by Harriet Martineau—the cheapest made from tin or even cardboard.[5] This diversity we interpret as being more than a matter of happenstance or fashion: it reflects the diversity of kinds and degrees of hearing loss that individuals experienced—bilateral, unilateral, high frequency, low frequency, as well as the range of aetiologies in disease, accident, inheritance or the ageing process (among others). More than that, certain kinds of hearing aids were also often specially designed for specific contexts and roles: church/chapel, parlour conversation, public gatherings, concerts and theatres, and even to match the social situation of the user: socially elevated, in mourning, senior in age—or disguised as clothing or other bodily accessories for the self-conscious. Suffice to say there was no such thing as the hearing aid, only many varieties thereof, and the range available increased in number as the century progressed.

In this chapter we look first at two of the leading mainstream hearing aid manufacturer/vendors—first the Rein company that was non-medical in orientation and expected direct visitations, whereas the medically oriented Hawksley very much expected mail order interaction instead. We then explore the more opportunist vendors studied through criticisms of their trade from such deaf journalists as George Frankland and pseudonymous 'Evan Yellon'. Finally we examine how hearing aids could be creatively adapted by their end users to show how far users could make devices their very own through contained adaptation.

1 Rein's 'Paradise for the Deaf'

Each of the major vendors of hearing aids in nineteenth century London highlighted their longevity of establishment and metropolitan location in their publicity: the Rein company was sometimes (mythically) dated back to 1800,[6] Arnold to 1819, and Hawksley to 1869. As Bennion emphasizes, Rein was the most prestigious with many of the most spectacular and expensive products associated with royal customers.[7] The high street emporium, Rein's 'Paradise for the Deaf', was located (up to 1916) at 108 The Strand—significantly central to the cultural and business life of the City. The prominence of this venue in London and the company's patronage from royalty were significant in attempts to attract customers. Other markers of social status were used too: in a dedicated publication of 1900 testimonials, evidence as to the 'efficacy' of Rein's devices was apparently available, notably from 'one of Her late Majesty's Judges', who used it in court.[8] Further credibility of Rein's 'Acoustic Repository' was drawn from winning prize medals for their hearing aids at international Exhibition. From such international recognition, Rein went so far as to claim to be 'the only Makers of real Acoustic Instruments for extreme, and every other degree of Deafness'.[9]

Unlike the other leading hearing aid companies, the Rein Company used patents and its patentee status to reinforce the impression of propriety in its instruments. In its centenary publication, the company emphasised that Rein had produced a succession of ten 'Patent Aurolese' devices, suggesting that this gave them a *longue durée* guarantee of quality. Evading any questions about whether any of these Aurolese devices bore any current patent,[10] Rein represented the series as the 'stepping stones to our present scientific results' as embodied in the Rein hearing aids of its centenary year. Acts of conspicuous consumption shaped the contours of many sales from this most famous of all the companies.[11] Those made by F.C. Rein, generally bore their trademark, a claim about patentee status and the company's location embossed as a mark of authenticity—implicitly a warning against counterfeiters.[12] In fact, the high quality materials used by Rein, including silver-plating, and the elaborately engraved and burnished finishes were a challenge to any aspiring counterfeiter or pirate. Rein's hearing devices were indeed designed to be the most upmarket available, with aesthetic qualities that the most affluent customers could conspicuously display in public.

Such were the Rein Company's successes both among the affluent and the more modest purchaser of tin and brass devices that no voices of disapproval can be found against the Rein Company or its products. In fact, one prominent turn of the century hard-of-hearing critic thoroughly approved of them, and recommended the Rein above all others. What mattered to this critic was the extraordinary tact of proprietor Frederick Charles Rein (born 1842) as well as the extraordinarily broad range of stock Rein held to meet any degree of hearing loss. Working under the pseudonym Evan Yellon, presumably to avoid libel proceedings against those he exposed as fraudsters, this journalistic 'Surdus' in search of authentic hearing enhancement conducted an interview with a Rein company representative c. 1909.[13] Yellon reported that Rein had many medals from International Exhibitions and diplomas, and 'sufficient testimonials to paper a barracks'. More than that, Rein's staff was 'most kind and exceedingly skilful', and deaf people were 'strongly recommended' to go to this company.

Recognising Rein explicitly as prime amongst those who sold 'Genuine Aids to the Deaf', this interview revealed Rein's primary commitment to selling hearing aids to those who came to visit the company's headquarters at 108 The Strand; this was to ensure that the particular needs of a customer's hearing loss were met by the most appropriate device. Rein conducted little mail order business since, like a top-calibre tailor, it set great store by personal testing, often for several hours, to ensure an appropriate fitting for customers:

> Very few people realise this, but it is the fact: you cannot pick up an aid off hand. We receive hundreds of orders by post, and in the cases of people residing abroad, or in the Colonies, we sometimes try to fit aids from particulars sent to us; but as a rule it is really necessary for the deaf person to come to us, [because] we can do most and be of most service then.[14]

Yellon emphasised this ethos of personal care and respect as a mark of Rein's non-exploitative character. After Rein had tested the ears, and given advice Yellon reported that 'no one is ever pressed to purchase an aid when calling at 108, Strand'.[15] As Rein did not advertise, its principal mode of promotion was thus through exhibition prizes and word of mouth. Even for those experiencing both vision and hearing problems, Rein was able to accommodate special spectacles, and also to use hearing aids while they chose especial glasses to suit their optical needs. Rein was not only very sensitive to the personal experiences of hearing loss, but also to the means of its customers. The range of products included everything from inexpensive trumpets for a few shillings suitable to the least affluent customer to more powerful instruments of up to 15 guineas for the well-to-do, and aural chairs of up to 100 guineas for the hard of hearing who wished to hold court in regal fashion. Whatever their degree of affluence, the same principles were applied to all of Rein's customers, and this earned Rein's company a great reputation for its products and service.

Rein was well-known for always taking care to avoid pressurising prospective customers. Indeed Rein was very reluctant to press on the hard of hearing any mere novelty such as the electrical aids that, since the 1880s, had arisen from adaptations of the telephonic microphone. Even after twenty years, the Rein company apparently found them in 1909 still 'too noisy' and still none beat the performance of their 'old instruments'. The company did not, however, rule out the prospect that the 'perfect electric aid' might yet come along one day.[16]

While Rein's business was premised upon personal care, what opportunities were there for individuals who could not visit in person? From 1869 a new alternative was open to them: the Hawksley Company based in 357 Oxford Street, London. Unlike Rein, this mail order rival made much of its medical credentials, implicitly tied to the rising clinical subspecialism of otology.

2 Hawksley's Otoacoustical Mail Order Business

Whereas the Rein Company relied almost entirely on word of mouth recommendations and personal visits, Thomas Hawksley's Company issued a mail order catalogue of 'Otoacoustical Instruments to Aid the Deaf' roughly once a decade. This implicitly complemented the Rein Company's older face to face business in the Strand, also to offer practical advice to potential customers. According to its third Catalogue of 1895, all devices were not only invented by Thomas Hawksley, but also made by himself at his Oxford Street premises. Hawksley's credibility was based on being Acoustical Instrument Maker to the 'Principal Aurists' in England, Scotland, and Ireland, France, Germany, India and the United States of America as well as to three major London hospitals: Middlesex, Guy's and St George's. In contrast to Rein, the Hawksley Company emphasised its medical connections as if to authenticate the 'oto-acoustical' credentials outlined above. Not only did Hawksley have close clinical connections, but it offered special purchasing terms for customers at Hospitals, Institutions, Charitable Societies, and Workhouses when application was made through 'medical officers'.[17] Moreover in early twentieth century editions of the Catalogue, Hawksley appended extracts from the lead otologist Sir William Dalby, on 'The Preservation of Hearing'; this signalled the ever closer relationship between the recently emerging subspecialism of Otology and Hawksley's Otoacoustical trade.[18] This adoption of medical alliances looks to be a threefold strategic move by Hawksley to enhance its credibility as trustworthy authority; to build up new clients and collaborations in the medical world and, relatedly, to distinguish its approach from that of its more established competitor, Rein.

To nurture demand for its products, the narratives of Hawksley's catalogues followed Martineau in pathologising hearing loss as if a social problem that had been created by deaf people's failure to acquire and use a hearing aid. A 'deaf person', so all of its catalogues averred, was 'always more or less a tax upon the kindness and forbearance of friends'. Hence it became the deaf person's duty to 'use any aid which will improve the hearing and the enjoyment of the utterances of others' without thereby engendering 'any murmuring about its size or appearance' among them. However, Hawksley did not solely attach blame to the hard of hearing as the source of social problems here: insensitivity among the hearing could also make communication more difficult. Indeed, he admitted that deaf people also had a 'just complaint' against many friends and public speakers who 'render their affliction apparently greater by an indistinct and mumbling utterance.' But that was the limit of bilateralism acknowledged by Hawksley in communication between the hearing and the hard of hearing. Overall, enhanced communications for deaf people were commodified into the purchase of a hearing aid.

Hawksley's catalogue emphasised that the appearance of hearing aids ought not to intrude upon social interactions. This brought to the force the issues of aesthetics also came to the fore in ways not raised by Martineau:

> the ingenuity and taste of the instrument maker are required to construct mechanical aids to hearing which shall combine gracefulness of form and appearance without detracting from their efficiency, for the burden of deafness is great and the sensitiveness of the sufferers should not be wounded by the necessity of announcing their affliction to the public by having to use instruments either unsightly in form or objectionable in color or material.[19]

As for Rein, the most ornate Hawksley devices were made of polished brass, or sterling silver, decorative yet practical (e.g. collapsible), and wealthier users may well have owned several aids for use in different settings e.g. an India-rubber speaking tube for everyday conversation at home while adopting a decorated silver-plated dome for the Opera box.[20]

But in a further contrast to Rein, Hawksley did not expect that all customers would visit to have their hearing tested: Hawksley aimed for an international and imperial market, perhaps to complement Rein's appeal to a London-based clientele. The Hawksley catalogue invited prospective customers to try instruments out 'on approval' by writing in with 'full particulars of the nature and degree of deafness' that they had experienced[21] using standard testing techniques outlined in Hawksley's catalogue. For those who did not have access to the relevant kind of equipment, Hawksley would mail out to them a portable measuring device—the ABC acoumeter—that he had especially developed for prospective customers. From analysing their self-reported results, and upon receipt of a 'satisfactory London reference', Hawksley would then send an 'assortment' of potentially suitable assistive instruments for a five day free trial.

In contrast to Rein's bespoke approach of finding a suitable device for any personal visitor from his shop's encyclopaedic store, the Hawksley

1909 catalogue offered a standardised 11-fold classification to enable prospective purchasers to identify the kind of hearing aid that they most likely needed. These devices ranged in price from simplest devices reflecting the complexity of design and size from £0.7.6 for those of frugal means with more expensive devices ranging up to a dozen guineas, albeit with none as expensive as Rein's.[22] Each device is characterised either in term of its mechanism or its likely category of user. There were conversation tubes for the 'extremely deaf' with sound collected directly from the lips of speakers; sound collected and conveyed via metal trumpets and horns; disguised aids such as hats, bonnets, canes; small pocket-portable devices with resonating chambers; bone conducting devices; reflective devices that acoustically focused sound as it entered the ears; table top or lap instruments that traded portability for amplification; loud-speaking battery-powered micro-telephones; multi-user instruments for churches, and chapels, and devices specifically constructed for educating deaf children via amplified oral methods.[23]

Among these models we find celebrity-named models borrowing from the example of both famous deaf people e.g. the Harriet Martineau and Joshua Reynolds forms,[24] and those named after medical practitioners, for example the Irish aurist H. McNaughton Jones and English otologist William Dalby.[25] Many devices were made by Hawksley in coordination with inventors such as McNaughton Jones while others were imported such as the electrical acousticon invented by the US engineer, Miller Reese Hutchison.[26]

The Hawksley Company, like Rein, had a very good reputation for treating its customers fairly. As Evan Yellon wrote in the first edition of *Surdus in Search of His Hearing* in 1906, the Hawksley company was one of the few that could be trusted on the new electrical devices 'I believe that they can show every form of aid, electrical or otherwise; they will certainly offer sound advice'.[27] Yellon was drawing an implicit contrast here to those other kinds of vendors whose advice was not so sound, and it is those to which we turn next.

3 The Opportunist Hearing Aid Vendors

Before the First World War at least, Rein and Hawksley did not advertise in the popular press: they were too successful to need that kind of publicity. However, in Victorian national, local and provincial press we can find many classified advertisements for aids to the deaf via mail order or travelling agents—far fewer than for short-sightedness or myopia. These advertisements for alleviation or even cure of deafness were typically for equipment, books, or for discreet appointments with an 'expert' advisor. The temptations to purchase via mail older were manifold for the embarrassed or nervous bearer of hearing loss who did not want to go to a high street vendor like Rein or Hawksley. The alternative vendors did not operate from high street premises bearing a large range of hearing aids for personalised fitting. Instead they typically made bold claims about the capacity of their generic devices to relieve any and all kinds of hearing loss. Unlike Rein and Hawksley, these opportunists did not offer money back guarantees if hearing aids proved unsuitable after a trial period. Although much criticised, such practices were entirely legal up to the early twentieth century under the 'caveat emptor' rubric: 'buyer beware'.

The hard of hearing community did not, however, accept this exploitation of their situation. This was the era of the new journalism in which writers for newspapers and magazines did not passively report on the world around them, but sought actively to expose crime and fraud.[28] Campaigning journalists such as Yellon and George Frankland (see Chap. 7) subjected the less scrupulous hearing aid vendors to public exposés. Yellon wrote in *Surdus in Search of His Hearing* (1906) of the many fraudulent companies that sold hearing aids by post. One of these fashioned himself as 'Professor Keith-Harvey' who advertised patented 'Aural batteries' heavily in popular magazines and journals claiming successful cures of (un-named) eminent patients. Yellon showed that whatever personal details were submitted by letter, the 'Professor' issued the same diagnosis and same course of battery therapy to all applicants by return of post. After further trials with the Keith-Harvey 'system' thus delivered, Yellon concluded that if any deaf person had found 'relief or cure' thereby, this could only have been by 'Faith' rather than any electrical agency.[29]

This critical response to disingenuous vendors reflected a broader trend of activist journalism in the pages of dedicated late nineteenth century newspapers such as the *Deaf Chronicle*. This publication claimed to represent all conditions of deafness, including the hard of hearing and their travails with exploitative 'cure' merchants. In its Capanbells column, readers regularly saw campaigns in the 1890s against the 'Quack Doctors who profess power to cure deafness': the column warned readers not to believe newspaper advertisements declaring 'Deafness Curable' with new ear gadgets. To supplement this in 1892 it reproduced the cautionary 'Swindles on Deaf People' by a 'partially deaf' journalist that had recently published in *Tit-Bits*. This journalist reported recently receiving a pamphlet for an 'artificial ear-drum' from one particularly aggressive company, which promised hearing restoration or a full refund. Upon enquiry he was soon advised that his particular deafness was curable by a gold-plated device at a cost of £2 11s 3d. A 3 month trial revealed this to be ineffective, so the correspondent asked for a refund: this prompted instead a demand for full payment with a County Court summons to follow if not forthcoming. Several of the journalists' neighbours reported that they had also experienced such hollow threats.[30]

This was just one of a series of similar episodes recounted by the journalist: 'How it is I don't know', but proprietors of similar companies 'have found out I am deaf' and sent him ever new pamphlets. We can infer that the companies involved in such mail order enterprises shared customer names and addresses. They could be confident thereby of a sustained clientele for their ever new varieties of ineffective devices from unhappy affluent hard of hearing people wanting 'cures'. But it was not just mail order companies that exercised such opportunism over those anxious to return to previous norms of conversational capacity. The campaigning journalist reported a company agent visiting a large town with a new device to offer. After a few questions, an ear inspection, and a check on whether he was in a position to pay £2 14s 6d, the agent inserted two instruments into the journalist's ears. When the agent simply told him that he could thus 'hear better' the journalist tried to check this with the traditional clinical test for degrees of deafness: finding the furthest distance at which a pocket watch could be heard ticking.[31] The agent physically prevented him applying this pocket watch test, maintaining that he should not expect to be able to hear it even with the new hearing aid. The suspicions of the investigative writer were thus vindicated and he departed without purchase.

We return to the role of the broader deaf community in monitoring the opportunist hearing-aid trade in Chap. 7. In the Sect. 4 we turn to the context of hearing aids in use, for at least some users of hearing aids were evidently satisfied enough with their purchases to deploy them regularly over long periods.

4 Personalising Hearing Aids in Use

Faced with opportunist exploitation, the partially deaf or hard of hearing had long used their own creativity to manage their communications. In a letter to *Blackwood's Edinburgh Magazine*, one partially deaf M.D.

of Aberdeen, Thomas Morison, reported in 1823 his improvement of the traditional hearing trumpet. Beyond his 'most sanguine expectation' this device had enabled conversation 'with the utmost ease to myself, and without exertion to the person addressing me'. Instead of seeking to profit, Morison advertised his technique openly:

> It is the establishment of the principle of this improvement upon the Ear-Trumpet to which I am solicitous to give publicity, leaving to younger men to make experiments upon the length and diameter of the tube, and of other parts of the instrument.[32]

More generally it is evident that much usage of hearing trumpets was creative and creative adaptation was indeed an integral part of a users' relationship over many years.[33] The life-course evolution of hearing aids fit this pattern of end-user creativity in ways that have not previously been observed by historians, more so by Museum curators. More than just artefacts of disguised assistance or of bodily augmentation, hearing aids were 'things' that circulated in everyday life and contributed to social status in ways well-established within Victorian material culture studies, subject to the characteristic forms of relationship between designers, users, and user-designers.[34] As Blandy and Bolin have recently discussed, cultural interpretation of artefacts (such as hearing aids) requires observation of the objects' weight, size, the materials used, trademarks, marks of wear or repair etc. It also entails inferential analysis of less obvious meanings such as (apparently) non-functional ornamentation.[35]

For example, looking at the hearing trumpets held by the Thackray Medical Museum, many are too delicate for display owing to extensive wear and tear; this of itself is evidence of these devices' long-term use and value to their users. A particularly fascinating example in the collection is an Arnold Hearing Horn Arnold-branded hearing horn in gunmetal[36]: (Fig. 2)

This was made of nickel-plated gunmetal and unadorned apart from the Arnold trademark. Its slight denting enables us to infer that it was used frequently. Moreover, its (presumptively) female user valued it enough to make/get made a draw-string bag to contain it: a personal adaptation, made of hand-sewn modest (possibly curtain) fabric decoratively embossed with flowers. Unlike their Hawksley competitors, which could be bought with leather, silk-lined carrying cases,[37] Arnold's London domes were not it seems normally purchased with bags. This Arnold example was far from

Fig. 2 Thackray Medical Museum collection, Object No. 2005.0338 © photo credit K. Sayer

being the only aid to have been adapted its user. Another London dome, a brass example made in France by Audios c. 1890, is similarly dented and covered in a close-fitting crochet cover (the cover being typical of 'local peasant crochet work'). A third device, a conversation tube with an ivory horn, c. 1890, has had tape placed by a user over the joins between tube and horn, tube and earpiece; this was presumably to protect those joins from wear and tear or fingers marks building up in use.[38]

If we consider the users of hearing aids, through their production of alternative and alteration of existing designs, they over-rode any control over the transaction presumed by the maker or vendor. Each modification by a user, opting in or out of what was pre-scribed at the point of sale, tells us about time spent in the care of the object, and adaptation for personal use. Victorian aids to hearing might involve dressing to hear, but were also selected for purpose and had to be fit to use: maintained both as an aid, and within the context of life as it was lived by the hard of hearing.

5 Conclusion

This chapter has focused on the opportunities open to 'hard of hearing' people to manage their condition by using hearing aids. We have seen how many purchase opportunities there were available, and how many varieties, reflecting the myriad kinds and degrees of deafness experienced. Customers of Rein and Hawksley might have expected to sell and then care for one well-matched assistive device at home; yet even they might have found a need for another kind of hearing to operate elsewhere. And as hearing deteriorated over time, they might need to purchase further hearing aids with more robust amplification—in contrast to lip-reading which worked (in suitable conditions) irrespective of the degree of hearing loss. Then again if they had been to an opportunist vendor with no money back guarantees, their repeated purchasing of hearing aids could have been even less satisfactory. Nevertheless, some at least in the hard of hearing world were satisfied with their hearing aids enough to preserve them for many years and nurture them as an essential accessory. And it was in this increasingly reciprocal world that medical perspectives were re-introduced, especially via the Hawksley Company's Oto-acoustical clientele. Indeed, a rising medical sub-group of otologists saw increasing prospects for re-engaging with the deaf in their collaboration both with hearing aid manufacturers. And it is to that medical group we turn in the Chap. 6.

Notes

1. As Jaipreet Virdi has noted there is significant visual evidence of historical photographs showing hearing aids in use, although not in paintings· see discussion in 'Using an Ear Trumpet' at http://fromthehandsofquacks.com/2013/01/18/using-an-ear-trumpet/ (last accessed 16th Dec 2016).
2. Kenneth W. Berger, *The Hearing Aid: its operation and development* (Michigan, USA: National Hearing Aid Society, 1970, revised 1974); Elisabeth Bennion, *Antique Hearing Devices,* (Vernier Press: London & Brighton, 1994).
3. See Julie Anderson, 'Report on the Thackray Hearing Aid Collection Thackray Medical Museum, Leeds' (Thackray Museum, unpublished report, University of Kent, 2010).
4. Some of these devices were made out of very simple inexpensive materials, such as tin, or even cardboard; the latter are described in the Hawksley catalogues, though understandably no surviving example has

yet been found. Other devices were fabricated from seashell—and later faux shell—possibly because some of the earliest aids were actually modified seashells. Through this material the association remained strong, as well as shell being decorative and reputedly low in vibration for the user. John Bell & Croyden, London, Wigmore Street, W1, *The complete hearing service for the deaf*, paper pamphlet, 23 p., 21.5 × 13.4 cm (London: privately published c. 1930). http://johnjohnson.chadwyck.co.uk/pdf/tmp_6472252476951592887.pdf (last accessed online 4th Nov 2015), p. 12.

5. For cheap materials used in making hearing trumpets see Charles James Blasius Williams, *Memoirs of life and work* (London: Smith, Elder, & co, 1884), 433.
6. For evidence debunking the of-repeated myth that the Rein company was founded in 1800, see Dominic Stiles 'Acoustic instrument makers in the Strand, acoustic "throne" myths, & Frederick Charles Rein & Son', http://blogs.ucl.ac.uk/library-rnid/2016/11/11/acoustic-instrument-makers-in-the-strand-acoustic-thrones-frederick-charles-rein-son (Last accessed 30th March 2017). Charles Frederick Rein is first listed as a Surgical Instrument Maker in *The Medical Times and Gazette*, 1855 Volume 2; Volume 11, 308—presumably the father Charles F. Rein. Bennion p. 329 lists the Rein Company as starting in 1851 at The Strand. F. Charles Rein, Aurist and Acoustic Instrument Maker, 108 Strand, London; then Rein and Son, 1865–1866 and then Frederick C. Rein 1867–1917 all at the same address. The first Rein patent was taken out by one of the two Frederick Charles Reins (father and son) who operated as Acoustical Instrument Maker': this was 'Apparatus for Excluding Sound from the Ear', British Patent 1864, 1st December, No 3000'. We thank Dominic Stiles for pointing out that in 1900, following the death of the younger Rein, Rein senior's widow apparently sold the family business to a neighbouring Optician, Charles Kahn. Evidently Kahn continued to use the brand Rein name in the 1900 promotion pamphlet discussed below, in which the 1800 foundation date was concocted. Khan even adopted Rein's persona in interviews with Evan Yellon, *Surdus in Search of His Hearing* (London: Celtic Press, 1906).
7. Bennion mentions Queen Victoria and King Goa VI of Portugal as customers among the world's royalty, *Antique Hearing Devices* 21, 34.
8. F.C. Rein & Son, 'The Paradise for the Deaf,' Centenary promotional pamphlet, 1900. F.C. Rein Publicity pamphlet reproduced in Mary Lou Koelkebeck, Donald Calvert and Colleen Detjen, *Historic devices for hearing: The CID-Goldstein collection*, (St. Louis, MO: Central Institute for the Deaf, 1984). For more on Rein see Berger, *The Hearing Aid*, 234–235. For a European-style 'flesh-coloured' flexible hearing tube

of the period see http://www.sciencemuseum.org.uk/broughttolife/objects/display.aspx?id=92088 (last accessed 11.10.13); also Deafness in Disguise, http://beckerexhibits.wustl.edu/did/index.htm curated by the Washington School of Medicine and Bernard Becker Medical Library (last accessed 11.10.13).

9. See discussion of the Rein display at 'The Paris Exhibition', *The Lancet*, 112 (1878), 136–137.
10. Similar models made by e.g. Hawksley can be found on the site 'Deafness in Disguise' http://beckerexhibits.wustl.edu/did/advert/part2.htm (last accessed 9th October 2015), and by Maw & Son in Bennion, *Antique hearing devices*, 27.
11. An undated Rein notebook in the Thackray Museum's library collection shows some evidence of personalised home fitting service to the more famous hard of hearing customer seeking discretion (including William Ewart Gladstone). Our thanks to the Thackray Librarian, Allan Humphries, for showing us this document.
12. For patents see Kenneth W. Berger, *The Hearing Aid: its operation and development* (Michigan: National Hearing Aid Society, USA, 1st edn. 1970, revised 1974) and Graeme Gooday and Karen Sayer, 'Purchase, Use And Adaptation: Interpreting 'Patented' Aids To The Deaf In Victorian Britain' in Claire Jones (ed.) *Rethinking Modern Prostheses in Anglo-American Commodity Cultures, 1820–1939* (Manchester: Manchester University Press, 2017) 27–47.
13. 'Mr. F.C. Rein's Aids for the Deaf' in Evan Yellon, *Surdus in Search of His Hearing* (London: Celtic Press, 1910, 2nd edition), 55–58.
14. *Surdus 2nd edition*, 58.
15. *Surdus 2nd edition*, 60–61.
16. *Surdus 2nd edition*, 60.
17. T. Hawksley, *Catalogue of Otacoustical Instruments to Aid the Deaf*, (London: John Bales Sons & Danielsson, 1909), 6th edition, 2.
18. William Dalby, 'The Preservation of Hearing,' *Longman's Magazine*, 32 (1898): 218–226, included as unpaginated end matter in the 1909 catalogue. Dalby was Consulting Aural Surgeon to St George's Hospital and in 1899–1901 served as the first President of the Otological Society of the United Kingdom.
19. T. Hawksley, *Catalogue of Otacoustical Instruments to Aid the Deaf*, 1883 3rd edition, preface.
20. Myk Briggs, photographer and independent collector, email to Karen Sayer 14.10.13; 'As for which to choose, it is a minefield of vendor opportunism … More work needs to be done if manufacturers' sales/distribution records can be located how many such devices were made and sold, and to whom, or indeed how many patented aids to hearing were

sold in proportion to unpatented devices, and whether this changed over time/as any patents were defended.'

21. Hawksley, *Catalogue*, 10. The section headed 'Measurement of Hearing' referred to Hughes' Electric sonometer, Differential Acoumeter, Politzer's acoumeter, Galton's whistle, ABC Acoumeter. The hard-of-hearing journalist George Frankland received his copy of the Hawksley 1909 catalogue in request for information about the company's specialist equipment for teaching the deaf. This is preserved in the Action on Hearing Loss Library with an accompanying letter inserted opposite page 272 dated 6th April 1911: 'Dear Sir, In reply to yours, we send you our 1909 catalogue & on p. 72 you will find a description of the kind of apparatus we supply for teaching the Deaf. We shall be very pleased to send you one for a two weeks trial, free of charge. The power of the instrument can be increased by making the tube larger in diameter. Yrs faithfully, Hawksley and Son'.
22. See Hawksley, *Catalogue*, 'Class B (series 1) Simple Rigid Cones, 26; also Arnolds catalogue, 341, Fig. 1161, & Allen & Hanburys 1923, 190, No. 21153.
23. Hawksley, *Catalogue*, 20 Classifications of instruments.
24. Hawksley, *Catalogue*, 32–33. Also referenced are Dalby, 46, and the Rawlins resonator, 54 for judges to use under their cloaks.
25. The 'H. McNaughton Jones' model listed a portable cone device in Hawksley *Catalogue*, p. 35, is quite distinct from the device shown in H. McNaughton Jones, M.D. 'An Acoustic Aid for Persons Partially Deaf From various Causes' *Proceedings of the Royal Society of Medicine* (Otology section) 3 (1910) 40.
26. For this and other electrical hearing aids see Hawksley, *Catalogue*, 69–71.
27. Evan Yellon, *Surdus in Search of His Hearing* London, 1906 (1st edition).
28. For the fearless 'New Journalism' trend for exposing crime and exploitation in the 1890s e.g. W.T. Stead, see Laurel Brake & James Mussell, Introduction. 19*: Interdisciplinary Studies in the Long Nineteenth Century*, 16 (2013). doi:http://doi.org/10.16995/ntn.669.
29. Yellon, *Surdus in Search of His Hearing*, 16–21.
30. 'Swindles on Deaf People', *The Deaf Chronicle*, 1892, 142–143.
31. This pocket watch technique as used by James Kerr of Bradford for testing hearing is discussed in Chap. 6.
32. Thomas Morison, T. 'New Ear Trumpet,' *Blackwood's Edinburgh Magazine, 14*(79), 199.
33. For studies of user discretion in the history of technology, see Nellie Oudshoorn & Trevor Pinch, *How Users Matter* (Cambridge, MA: MIT Press, 2004).

34. Jennifer Sattaur, 'Thinking Objectively: an Overview of 'Thing Theory' in Victorian Studies', *Journal of Victorian Literature and Culture*, 40 (2012). 347–357; Owens, Alastair, Nigel Jeffries, Karen Wehner, and Rupert Featherby. 'Fragments of the Modern City: Material Culture and the Rhythms of Everyday Life in Victorian London', *Journal Of Victorian Culture*, 15(2) (2010): 212–225; Marieke M.A. Hendriksen 'Consumer Culture, Self-prescription, and Status: Nineteenth-Century Medicine Chests in the Royal Navy', *Journal of Victorian Culture*, 20 (2015), 147–167.
35. See Doug Blandy & Paul E. Bolin, 'Looking at, Engaging More: Approaches for Investigating Material Culture' *Art Education*, (July 2012), 40–46, 41–44.
36. Thackray Medical Museum collection, Object No. 2005.0338 © photographic image, Karen Sayer.
37. For an example see Alex Peck, Medical Antiques, 'A late Nineteenth Century large London Dome ear trumpet by Hawksley in its original leather and silk lined carrying case' Image © 2013 Phisic at http://phisick.com/item/london-dome-ear-trumpet-by-hawksley/ (last accessed 07.10.13).
38. 'It is French and I am told very typical of local peasant crochet work, so I believe it to be a home-made addition', Email to Karen Sayer from Myk Briggs [bigloaf@gmail.com] (14.10.13) http://www.eartrumpets.co.uk/pictures2.php?pageno=23&buyno=128 (last accessed 14th October 2013); 'one of my conversation tubes (#171) has home made sleeves at either end—they are not as manufactured but definitely of the period—I think to catch the finger traces off the user' Myk Briggs email to Karen Sayer [bigloaf@gmail.com] (14.10.13) http://www.eartrumpets.co.uk/trumpets5b.php?ref=194&picno=D3x6770.jpg&field=&fieldno=0&test=&sort=buyno (last accessed 14th October 2013).

CHAPTER 6

Preventing Deafness: Two Medical Approaches

Abstract The cultural climate of early twentieth century Britain combined eugenic and economic factors which threatened the well-being and status of all kinds of deafened people. In this context we discuss two otologists' attempts to (re)medicalise deafness with a new emphasis on how to prevent rather than cure it. We look at Percival MacLeod Yearsley of London and his eugenic obsession with the small minority for which deafness was allegedly heritable and the Glasgow-based James Kerr Love who aimed to show that most acquired deafness arose from health factors relating to poverty.

Keywords Otology · Specialisation in prevention of deafness · Economic concerns · Eugenic concerns

The pressures on hard of hearing people in Victorian Britain to find a 'solution' to their variant condition went beyond decisions on whether to communicate using a hearing aid, take lip-reading classes or correspond via friendly networks. Such choices were only meaningfully available to the affluent middle classes who could afford to pay for such options. Conversely for those who worked in Britain's factories, temporary or permanent loss of hearing did not necessarily need an extrinsic 'solution': many workers simply learned lip-reading as a matter of course to communicate across noisy machines on factory shop floors.[1] The key broader change was that the political and cultural climate at the turn

G. Gooday and K. Sayer, *Managing the Experience of Hearing Loss in Britain, 1830–1930*, DOI 10.1057/978-1-137-40686-6_6

of the twentieth century was introducing new forms of stigmatization against all kinds of hearing loss and deafness. Both industrial imperatives for uniformity and eugenic concerns about the quality of 'national stock' raised expectations that all kinds of deaf people should adapt to the hearing world's oral norms or face marginalization in unemployment.

Tolerance of variation in capacities for hearing reached its nadir in the decade or so preceding the First World War with eugenic projects to breed certain groups of deaf people–and others with variant heredity—out of humanity.[2] Closely related, were educational- institutional developments to promote oralist lip-reading culture for deaf children that had significant (if not entirely foreseen) implications for deaf adults. The Milan Congress on Deaf Education in 1880 had voted to eliminate sign language in teaching Deaf children—with only France and the USA voting against; hence the UK's Royal Commission on the Blind and Deaf in 1889 recommended compulsory oralist education.[3] Accompanying this approach was a new scheme of inspections by physicians with aural specialism to scrutinize children's hearing capacities and evaluate them as hearing, hard of hearing or deaf. The underpinning expectation was that value for money could be secured for the taxpayer that had to foot the bill for dedicated publicly funded oralist schools. A new generation of otologists was thus empowered to treat deafness as a species of economic problem to be minimized by preventing childhood deafness that arose through disease or infection.

This chapter investigates these broader contextual issues, focusing on two contrasting medical perspectives on the causes and management of deafness. We look at the rising medical sub-profession of otology that succeeded aural surgery as the principal medical engagement with deafness in the later nineteenth century. Whereas aural surgery had aimed to cure deafness, otologists' less ambitious aims were to *prevent* deafness, to investigate methods to eliminate deafness, and treat diseases of the ear (whether or not relating to hearing loss). We discuss two otologists' approaches to (re)medicalise deafness through a process of paternalistic management. The Otologist to London County Council, Percival MacLeod Yearsley, pursued a prophylactic project which focused for eugenic reasons on the small minority of cases for which deafness was allegedly heritable. By contrast the Glasgow-based James Kerr Love aimed to show that the large majority of cases of deafness arose from circumstantial health factors relating to poverty. While both otologists were sympathetic to the cases of deaf adults, before the First World War at least, neither saw it as their professional prerogative to offer them practical assistance. As we shall see, neither of these two forms of otological practice impressed journalists in the broader deaf community.

1 Bodily Diversity and Technological Challenges

Early in the nineteenth century, institutional debates on managing deafness were strongly related to religious concerns for the longer term welfare of deaf people.[4] There was also a rather more secular concern for the maintenance of verbal cultures in employment: factory owners and large-scale businesses were concerned that those with 'non-standard' bodily capacities should function effectively in ever-more intensive industrial-scale processes of production and commerce. For example, if employees could not hear verbal instructions or warnings, prospective employers could raise concerns about their propensity to operate safely and efficiently in a presumptively oral workplace. As disability historians have observed, such concerns led to the marginalization from the workplace of people with partial or complete limb-loss, blindness and or deafness, and many were institutionalized in asylums, hospitals, schools and workhouses across Britain.[5]

During periods of high unemployment those with any kind or degree of alleged 'disability' could find it hard to secure work unless they had either a sympathetic employer or could find a way of disguising their condition to 'pass' as able-bodied.[6] As Deaf historians have noted, this became an issue of particular pressure after the National Insurance Act of 1911 obliged employers to invest in unemployment funds for workers: accordingly employers became yet more selective about whom they would take on.[7] The capitalist imperative to increase the 'efficiency' of human labour increasingly expected workers to submit to larger-scale rationalized schemes of production that made little allowance for bodily or psychological variation.[8] That very same year it was such increasing challenges to securing employment that prompted Leo Bonn, a wealthy banker who lost much his hearing in later life, to set up the National Bureau for the Encouragement of the Welfare of the Deaf to help that broad constituency.

These broader trends affected the Deaf/hard of hearing adult in the less affluent sectors of the population by increasing their vulnerability to changing socio-economic circumstances. These were exacerbated following the international -wide shift to oralism for deaf children's education in the last two decades of the nineteenth century hastened by the Milan

Congress (see above).[9] Ironically this move to visual methods by teachers of both sign-language and lip-reading came at just the moment when a new technology problematized reliance on visual information. This was the rise of the telephone from 1876 as the first entirely non-visual modern means of communication.[10] It was clearly impossible to communicate with the telephone using skills in signing or lip-reading. Thus it was soon turned into a device for testing children's aural capacities for understanding conversation, excluding any recourse by them to visual clues from lips or fingers. While this helped to overcome the practical problems of testing hearing alone, for those unfamiliar with a telephone it was understandably a confusing ordeal, and thus not always successful.

In 1897 for instance, James Kerr as the Medical Officer, Bradford Royal Infirmary and Medical Superintendent to the Bradford School Board, observed to the Royal Statistical Society, that current tests of hearing capacity using tuning forks and whispered conversation were 'very unsatisfactory'. This was largely because the test subject could see what the test involved, thus responding with results in ways that did not exclusively rely upon hearing faculties; moreover, the human contribution to the test was not necessarily constant. Ideally, Kerr suggested a technologically standardized test with the subject hearing sounds of various graded volumes through a telephone, without any visual connection to the origin of the sound. Secondly these sounds would be made mechanically consistent by using a standardized interrupted current on a phonograph. While this approach promised in theory to remove uncertainties in the results, Kerr found it had not been found 'practicable'. Presumably this was because either the machinery did not work reliably or that the sheer alien-ness of the tests prompted unusable responses from the test subjects. Recourse was made instead to older methods of testing 'acuity of hearing' for the sound of a watch moving across a range of distances, pinpointing at what distance the sound was inaudible.[11]

As Esmail has noted, by the early twentieth century the increasingly widespread usage of the telephone further alienated or disempowered Deaf and hard of hearing people.[12] This was because the telephone normalized 'hearing' in a technologized fashion as the capacity to engage in a telephone conversation: a task which presumed no loss of hearing. When female telephone operatives were thus chosen for the Post Office switchboards, a clear exclusion was made for any prospective candidates who did not have 'good' hearing, verbal 'articulation' or self-possession. These requirements, along with an exclusion of any candidate who showed a strong local dialect, nervousness, 'hysteria' or 'defect of speech', would have excluded most applicants with the mildest of hearing loss—along with many others.[13] Only after the First World War did the General Post Office attempt to develop any telephone service for the hard of hearing—in response to complaints from deafened war veterans expecting to be able to use telephones as part of everyday business transactions.[14] Meanwhile the 'stone deaf' and non-speaking Deaf were completely marginalised by this mode of communication, and no facilities for them were available until the US Deaf community helped to create teletype services more than half a century later.[15]

Hard of hearing people's job prospects were considerably diminished by any challenges they encountered in using a telephone during their employment. The Bureau's challenge to find employment for Deaf or deafened people was made harder still with the rise of broadcast radio in the 1920s unless listeners had personalized earphones. While greatly benefitting blind people as much as the telephone in de-privileging the visual element of communication,[16] radio further removed the partially or non-hearing from everyday conversations. This expected capacity for hearing-based communication would challenge many in decades to come, especially as expectations on that front converged with economic and eugenic imperatives that stigmatized such hearing loss in the early twentieth century. An investigation of those imperatives is at the heart of this chapter since they explain why so many of the hard of hearing sought to disguise their condition and instead 'pass' as hearing by using hearing aids. Ironically again, one major source of amplification for hearing aid technology was the microphone or valve amplifier borrowed from telecommunications–the very devices that problematized the auditory capacities of the deafened in the first place.

While hearing aid manufacturers profited from the ever greater stigma attached to hearing loss in the early twentieth century, what were physicians doing to assist the varieties of deaf people? Rather than seeking modes to assist them, the main goal of otologists was to anatomise, diagnose and eliminate hearing loss.

2 Percival MacLeod Yearsley: The Eugenic Perspective

Supporting early twentieth century economic pressures to remove 'inefficient' forms of diverse hearing from the workplace was the practice of eugenics, specifically campaigns to eliminate any inherited tendency to alleged 'deficiency'. That at least was the evaluative language used to describe variations from the bodily 'norms' that business producers and military personnel alike imagined should constrain recruitment procedures. As historians of eugenics have observed, it was the British Army's difficulties in recruiting soldiers fit to serve in the Second Boer war of 1899–1901 that prompted a panic about national 'degeneration' in human breeding stock. The measures to enhance 'National Efficiency' included both better nutrition and the elimination of inconvenient inherited traits. Thus it was that concerns about the bodily capability of male workers for conscription in the armed services converged with eugenic concerns for 'efficiency' to discourage Deaf people from inter-marrying.[17] Professions that could assist in this eugenic project were thus tacitly granted approval to explore any useful facets of eliminating inherited bodily limitations.

In the early twentieth century a few eugenicists within the medical profession researched hereditary deafness to illustrate their Mendelian theories of inheritance. They sought thereby to direct social policy to focus on biologically heritable factors rather than any social, cultural, or epidemiological preconditions of health and welfare. Some medical practitioners sought the sterilization of deaf mute people to remove any reproductive capacity, as recommended by Alexander Graham Bell in his 1884 Memoir on the prospect of a new 'Deaf-Mute' race created by intermarriage between deaf-mute people.[18] The Bradford Medical Examiner (Kerr) cited above for example, reiterated Bell's eugenic concern, claiming that 'heredity is so strong in the production of deaf-mutism'. However, unlike Bell, Kerr argued not for sterilization to avoid this putative economic burden, but of gender-segregation and the imposition of oral methods as a method of social 'hygiene'.[19] While this mainstreaming of quasi-eugenic views created greater mistrust of medical professionals among the Deaf/hearing loss community,[20] it certainly served the purpose of civic administration for Kerr as concurrently the Medical Officer (Education) for London County Council.

One physician that more directly adopted Bell's eugenic perspective was Percival MacLeod Yearsley (1867–1951).[21] Early in his career Yearsley secured a major position as senior surgeon to the Royal Ear Hospital in London as well as Lecturer and Examiner to London's Training Colleges for Teachers of the Deaf. By 1908 he was also working under Kerr as the Senior Aural Surgeon (later otologist) and Medical Inspector for the London County Deaf Schools. In this role Yearsley reported to London County Council on how much the capital's rate payers spent on Deaf and hard of hearing school children. He thereby secured access to a large number of subjects for his eugenic research, categorising his subjects into

'Deaf', 'semi-Deaf' and 'doubtful' to help him pinpoint inherited causes of deafness in the first of these categories. Although this period of eugenic research was discreetly obscured in obituaries of Yearsley—for in later life he apparently mellowed to secure the 'respect and admiration' of otologists—his contemporaries certainly regarded him as somewhat of a 'stormy petrel' during the eugenic debates c.1911. After all, Yearsley published this research on the 'Condition of Hearing' not only for a medical audience in *The Lancet* but also more belligerently in the *Eugenics Review*.[22]

Between spring 1911 and summer 1912 Yearsley wrote a series of articles for *The Lancet* on 'The Education of The Deaf: Its Present State In England, With Suggestions As To Its Future Modifications And Development.' Even though Yearsley acknowledged that in 97% of cases deafness was 'acquired' after birth, his main interest was the 3% of allegedly congenital deafness. His conclusion to the three-part version of his *The Lancet* paper argued that while it would be impossible to eradicate sporadic cases of children with heritable hearing 'defect', he could use eugenic principles to 'enormously' reduce the number with congenital deafness that had to be educated at the expense of the tax-payer by preventing marriage among 'Deaf mutes', close family, alcoholics, syphilitic, and 'those with a family taint of insanity'. He concluded that out of 691 deaf children inspected for London County Council schools, a total of 284 (41%) were congenital cases. By eliminating these, he argued that more attention could be paid to the educational needs of the 407 cases of children with acquired deafness.[23] Indeed, Yearsley argued that physicians should lead the way: 'It is the medical practitioner who should be foremost in eugenics.'

This is confirmed in his closely related piece 'Eugenics and Congenital Deaf-Mutism' that he published concurrently in the *Eugenics Review* in which he drew more explicit conclusions about the 284 'undoubted congenital cases'. To fellow eugenicists he spelled out his master plan for the congenital deaf-mute 'I am afraid that this statement is a bold one, but I do not fear to say it here': sterilisation.[24] In economic terms the cost of educating each Deaf-mute in was £31 7s. 6d. per annum compared to £5 3s for the 'normal' child. This differential in cost was likely, he felt, to make his eugenic solution appeal to the London ratepayer.[25]

And it was not only the ratepayer who was supposedly to benefit from this eugenic project. When Yearsley finally got round to writing on 'Acquired deafness' and its causes in *The Lancet* in 1912, he planned for a future medical profession in which physicians would be educated to be responsible for this larger group of deaf people. Rather than discuss their experiences, however, he listed his comprehensive analysis of the broad range of health conditions that could precipitate post-natal deafness.[26] With respect to these causes Yearsley had little to say concerning the prevalence of disease among the city poor except that such was to be addressed by the sociologist and doctor. But he did at least reveal that in dealing with the causation of acquired deafness, he was 'working upon surer ground' in plans to prevent its occurrence: 'the prevention of acquired educational deafness embraces a number of factors. It means, among other things, better care of children generally, better hygiene, better feeding, better surroundings.' This environmental rather than genetic interpretation of acquired deafness is one that he held in common with other otologists who were less eugenically inclined.

Overall, however, Yearsley appears to have been more interested in enhancing the professional prestige of otology than care for the deafened. And it was in the growing task of the prevention of acquired deafness that his specialism's purported role lay:

> The prevention of deafness (in which lies, I am sure, the future of otology) needs a universal knowledge on the part of the profession of the etiology (sic) and treatment the causes leading to defective hearing, and such a universal knowledge can only be made possible by the cognition of otology as an important part of the syllabus.[27]

It is valuable then to compare Yearsley's views with a counterpart in Glasgow, James Kerr Love, who agreed both on the incurability of hearing loss and the prophylactic imperative, yet adopted a different emphasis in studying the origins of acquired deafness.

3 James Kerr Love: The Sympathetic 'Prevention' of Deafness

James Kerr Love practised as an otologist in Glasgow after qualifying there in 1880. He soon developed a research project into the limits of hearing and the acoustics of musical sounds; in recognition of this he was employed among other duties as an Aurist by the Glasgow Institute to the Deaf and Dumb, and aural surgeon (an historically inherited title) to the Royal Infirmary. After specialist study of the 'deaf-mute' in a work of

1893, two works in the early twentieth century sealed his national reputation in the UK: *Diseases of the Ear for Practitioners and Students of Medicine* (1904) and *The Deaf Child* (1911). However, it was not primarily these that are relevant to our story: Kerr Love had much sympathy for the experiences of deaf people, and unlike MacLeod Yearsley was not primarily concerned with prescriptions against inheritable forms of deafness. It was his transatlantic contact that gave him most sympathetic understanding.

Through his working with many varieties of deafness Kerr Love came to know Helen Keller who lost both sight and hearing in infancy (see discussion in Chap. 3) As one closely interested in understanding the perspectives of deaf people who had previously known hearing, he developed a close working rapport with this American deaf-blind icon.[28] Recognising her as an authority was Kerr Love's strategy for showing how proper respect for deaf people's views, especially in seeking to establish communication, was key to addressing their needs. Thus when the National Bureau for Promoting the General Welfare of the Deaf was launched in 1911, its instigator and main benefactor, Leo Bonn looked to Kerr Love as a sympathetic practitioner who could develop better public appreciation of the subject. Thus in 1912 Kerr Love delivered four lectures on *The Causes and Prevention of Deafness* (to an audience of clinicians, and leading figures (including teachers) from across the broadly constituted deaf community.[29]

In his prefatory remarks Kerr Love conceded that medical science had done little for the welfare of the deaf—a clear departure from the medical presumptions of William Wright uttered a century early (see Chap. 3). Indeed, an underlying theme of Kerr Love's lecture was that medicine *per se* could not hope to alleviate let alone 'cure' most cases of deafness/hearing loss.

> ….there is one reproach which is not often spoken of but is constantly felt by almost every otologist—he cannot cure old-standing deafness. In spite of all the advances of science, of all the triumphs of surgery, deafness of a few years' standing is seldom cured, usually gets worse, and the honest practitioner has to see his patient pass from one kind of quack to another, knowing all the time that the hopes hatched of the big promises of these rogues will be disappointed.

Kerr Love thus made plain his disgust for the opportunist advertisers for 'cures of deafness' that we discussed in the previous chapter. At this stage he expressed no interest in the prescription of hearing aids. Instead he focused on the 'fine field' that there was for preventing the acquisition of deafness.[30]

Since scarlet fever, meningitis and syphilis were the most common correlates of early acquired deafness, he argued that the only effective approach was prophylactic disease prevention via environmental and sexual hygiene. In focussing on unfavourable industrial and cultural working conditions which were a major correlate of acquired deafness, Kerr Love's approach differed from Yearsley in declining simplistic differentiations between 'congenital' and 'acquired forms' of deafness. Since these could be combined in many cases e.g. as the inherited tendency to later life hearing loss (e.g. otosclerosis) Kerr Love was thus very critical of Eugenicist approaches premised on such a specious distinction, and perhaps implicitly condemned much of MacLeod Yearsley's work:

> A new science of Eugenics has recently arisen, and the Eugenist asks himself the same question, and I regret to say sometimes answers it in the most empirical or pseudo-scientific manner. Sterilize the deaf, or make it illegal for them to marry, or shut them up in asylums, or fine them and imprison them if they have children. Such are some of the suggestions made. Some of the Eugenists are like doctors, who prescribe before they have made any study of the case. I did not call this quackery, I only called it empirical or pseudo-scientific, but it is very closely allied to quackery.[31]

Kerr Love was especially critical of those in the eugenic lobby who prejudicially correlated deafness with 'mental deficiency' as too often epitomised in pejorative construals of the category 'deaf and dumb'. While deafness might have problematized education more than blindness or lameness, 'deaf people are not more stupid than the blind or the lame.'[32]

As a further critique of eugenic approaches, Kerr Love devoted part of his first Bureau lecture 'The Nature and Causes of Deafness' to empathetically documenting the social challenges encountered by the many deaf people with whom he had worked. Chief among these was the sense of isolation often reported to him owing to difficulties in communication. This was compounded by the uncomprehending responses of fully hearing people who often assumed that communication by shouting loudly was helpful— if they thought the purported deaf person deserved attention at all. Generally, the 'hearing' response to deafness was much harsher than was that of 'seeing' people to the blind:

> We shout to the deaf man, the worst thing we can do by the way, and feel annoyed at the effort we have to make, but we watch the blind man, give him a hand over the crossing, and feel so much better for the slight effort. Anybody can understand a blind man; it takes a student to understand a deaf one.[33]

Worse than that, Kerr Love reported, people claiming to be deaf were too often suspected of fraudulent misrepresentation since their invisible condition offered no material evidence with which to win general sympathy. While the unemployed blind person 'with a placard or dog', the one-armed man with 'a street organ', the one legged man with his crutch could be 'sure of a penny' from a passing stranger 'if they care to pose for it', if a deaf man were to try begging for money he would be in 'danger of the police officer'. Similarly, in the court of law, a finger lost in the course of employment was sure of a 'sympathetic hearing' and financial recognition under the Workmen's Compensation Act of 1897, even if the worker concerned was still in employment. But not so the worker deafened by an accident at work or by the noise of environmental conditions: the worker 'who has lost half his hearing gets nothing' by way of compensation. Thus deaf adults and children alike were often unable to communicate their situation to achieve justice and philanthropic support: generally they were 'cut off from far more of the interest of life' than others who had lost bodily capacities.[34]

But for all of this sympathetic explanation of the social disadvantages that so often accrued to the less affluent deaf, Kerr Love's clinical analysis and prescriptions were no more about the emancipation of deafened people than it was about their eugenic elimination. Dedicated exclusively to medical 'science' this volume nowhere addressed the issues of communication by signing, lip-reading or hearing assistance that preoccupied many hard of hearing people. While critical of Bell's eugenic over-simplifications and cautious about accepting Mendelian interpretations of inherited hearing loss, Kerr Love's motivation to deliver the four invited lectures in winter 1912–1913, was very much as their sponsor Leo Bonn put it in the published 127-page shilling pamphlet version of the lectures: to make them accessible to all who sympathised with the 'great movement for the prevention of deafness'. Since Leo Bonn's philanthropic motivation related to his own late-onset hearing loss (no explanation of which was offered by Kerr Love) we can understand this as Bonn seeking to ensure that 'wherein prevention is humanly possible', deafness shall be 'far less frequent' than hitherto. Characteristically of the pre-war era, Bonn argued that this would not only add to the 'sum total of the world's happiness' but also 'increase the economic well-being of the community.' Thus again we see the financial rationale for minimizing the prospects of children becoming deaf.[35]

This would explain the polite but muted response given to the final publication of Kerr Love's lectures in the *British Deaf Times* in August 1913. This had previously published journalistic summaries of each of the lectures when delivered. Reviewing the volume, the editor by framing the most important issue as the 'grave problems' involved in the 'affliction of deafness, ameliorative and curative' which 'force[d]' themselves on the attention of 'all those who are concerned with the welfare of the sufferers'. In comparison to the multitude of works of teachers and 'missioners' dedicated to that broader welfare project, Kerr Love's was a 'deeper' enterprise that would 'add to the sum of human knowledge of an extremely abstruse subject'. Situating the *British Deaf Times* as 'non-medical' journal, it declined to criticise the treatment of 'highly technical matters' that were outside its 'domain' of endeavour.[36]

However, during the First World War Kerr Love served in the 4th Scottish General Hospital at the rank of major. He thereby saw a significant cause of deafness not covered in this 1913 volume: explosive damage to eardrums. This prompted him and others in the medical profession in turn to more ameliorative measures for state support for training in lip-reading and hearing aids. In the meantime, as Daniel Kevles had pointed out, the growing body of neo-Mendelian geneticists dropped deafness from the range of bodily conditions to be appropriated as ammunition for eugenics. Not least among the reasons for this were that the majority of deaf or hard of hearing children were born of hearing parents.[37] Thus closed one particular avenue of enquiry to re-medicalise deafness.[38]

4 Conclusion

We have seen how the experience of hearing loss and deafness became increasingly fraught by the early twentieth century as technological, economic and eugenic factors situated the hard of hearing in a world ever more expectant of normalised hearing capacity. While this was recognised by the medical professions, they were not well placed to assist adults with hearing loss. The appropriate approach to deafness was the subject of some debate among specialists in otological matters, and their resolution was primarily that this was a managerial issue that could only be handled for children, and from the perspective of preventing rather than curing or mitigating hearing loss. We have seen how Percival Macleod Yearsley in London focussed on hearing loss as an inheritable biomedical misfortune—whether through birth or of ill-health. James Kerr Love in Glasgow responded by pointing more emphatically to poverty, squalor and lack of education as the main causes of deafness to be ameliorated in a more empathic understanding of the deaf and hard of hearing. While they differed on the primary origins of hearing loss, both agreed that it was an economic problem, ideally to be prevented or eliminated. In the next chapter, however, we shall see how both Kerr Love and MacLeod Yearsley changed the tone and direction of their 'medical' approach to hearing loss in consequence of their experiences in the First World War—looking away from the circumstantial aetiology of towards state-sponsored evaluation of hearing loss and provision for its care.

Notes

1. For discussion of the high proportion of workers deafened by working among Lancashire looms see Ron Freethy *Memories of the Lancashire Cotton Mills* (Newbury Berks: Countryside Books, 2008, 104). The use of exaggerated lip-movements in 'Mee Maw' communication is illustrated here https://www.youtube.com/watch?v=TJVVqNO8UoI (12th Dec 2016). For further discussion of long-term deafness engendered by noisy factory work: Karin Bijsterveld, 'Listening to Machines: Industrial Noise, Hearing Loss and the Cultural Meaning of Sound,' in J. Sterne (ed.), *The Sound Studies Reader* (New York/London: Routledge, 2012) 152–167. Thanks to Janet Greenlees for advance sight of her piece, "The noise

were horrendous': The ignored industrial disease' in *When the Air became Important: A Social History of the Working Environment in New England and Lancashire, 1860–1960* (New Brunswick: Rutgers University Press, forthcoming).

2. Jan Branson, Don Miller, *Damned for their difference: the cultural construction of deaf people as 'disabled': a sociological history.* (Washington D.C.: Gallaudet University Press, 2003).
3. The 1893 Elementary Education Act in fact required local authorities to provide school education for impoverished blind and deaf children up to the age of 16 Jackson, *Britain's Deaf Heritage*, 122–123. Anne Borsay, *Disability and Social Policy in Britain since 1750* (Basingstoke: Palgrave, 2005), 106–110.
4. Christian organisations in particular raised theological concerns about whether eternal salvation was possible for those who might not (fully) comprehend redemptive speech. Neil Pemberton, 'Deafness and Holiness: Home Missions, Deaf Congregations, and Natural Language 1860–1890,' *Victorian Review* 35: 2 (2009): 65–82.
5. Michael Oliver & Colin Barnes, *The New Politics of Disablement,* (Basingstoke: Palgrave Macmillan, 2004) 2nd edition, especially Chap. 3, 'The rise of disabling capitalism,' 52–73. Anne Borsay, *Disability and Social Policy in Britain since 1750.*
6. Jeffrey A Brune & Daniel J. Wilson, *Disability and Passing: Blurring the Lines of Identity,* (Philadelphia, PA: Temple University Press, 2013).
7. Peter Jackson, *Britain's Deaf Heritage* (Edinburgh: Pentland Press, 1990), 188. Jackson speculates that the rise of 'dangerous trades' legislation might also have prompted employers to hire fewer deaf/hard of hearing employees to minimize their liability for accidental injury in the workplace.
8. This is sometimes later referred to as Taylorism, for which see David Noble, *Forces of Production; A Social History of Industrial Automation* (New York: Knopf, 1984).
9. Oralism was only significantly resisted in the USA (Gallaudet University alone focusing on sign language education), Jan Branson, Don Miller, *Damned for their difference: the cultural construction of deaf people as 'disabled': a sociological history.* (Washington D.C.: Gallaudet University Press, 2003).
10. For the disputed origins of the telephone c.1876, see Stathis Arapostathis and Graeme Gooday, *Patently Contestable: Electrical Technologies and Inventor Identities on Trial*, (Cambridge MA: MIT Press 2013): 87–106.
11. James Kerr, 'School Hygiene, in its Mental, Moral, and Physical Aspects, '*Journal of the Royal Statistical Society* 60 (1897): 613–680, 639. James Kerr (who died in 1940), like his near namesake James Kerr Love (1942)

originated in Scotland. The former was a Cambridge graduate and practised primarily in London and Bradford, ultimately specialising in ophthalmology. Dr. James Kerr British, *Journal of Ophthalmology* 25 (1941): 592–593. James Kerr Love practised primarily in Glasgow and specialised to his last days exclusively in deafness and hearing loss: 'James Kerr Love,' *The Lancet*, June 20 1942, 752.

12. Jennifer Esmail, *Reading Victorian Deafness: Signs and Sounds in Victorian Culture* (Athens, Ohio: Ohio University Press, 2013), 188–189.
13. G.H. Murray, 'Selection of Telephone Operators,' General Post Office London, 17th July 1901, BT Archives, London TCB 37/3. Our thanks to BT Archivist David Hay for sharing this source with us.
14. For discussion of how First World War Veterans with hearing loss increasingly demanded access to amplified telephones in the inter-war period, see Coreen McGuire The 'Deaf Subscriber' and the Shaping of the British Post Office's Amplified Telephones 1911–1939' (University of Leeds, unpublished Ph.D. thesis, 2016).
15. Harry G Lang, *A Phone of Our Own: the Deaf Insurrection Against Ma Bell.* (Washington D.C.: Gallaudet University Press, 2000).
16. The charitable British Wireless for the Blind Fund was set up in 1928 to provide radios especially adapted for blind users, the same year that the American Foundation for the Blind undertook the same role in the USA. See http://www.blind.org.uk/about/our-history/ (last accessed 11 December 2016).
17. G. R. Searle, *Eugenics and Politics in Britain, 1900–1914* (Leyden: Noordhoff International Publishing, 1976), 20–34.
18. 'If the laws of heredity that are known to hold in the case of animals also apply to man, the intermarriage of congenital deaf-mutes through a number of successive generations should result in the formation of a deaf variety of the human race.' Alexander Graham Bell, *Memoir upon the formation of a deaf variety of the human race.* (Washington, D.C.: National Academy of Sciences, 1884), 4. For broader discussion of Bell see Robert V. Bruce, *Bell: Alexander Graham Bell and the Conquest of Solitude* (London: Golancz, 1973), 409–420. Branson and Miller also cite an alternative version of the *Memoir* as Alexander Graham Bell, 'The formation of a Deaf Variety of the Human Race,' *American Annals of the Deaf and Dumb,* 29 (1884): 70–77; Branson and Miller, *Damned for Their Difference*, 261.
19. James Kerr, 'School Hygiene,' 674. For an example of hygiene language used in advertising courses of deaf education, see 'International Health Exhibition inc. 'Oral Instruction for Deaf and Dumb' in *Pall Mall Gazette,* October 21 1884.

20. Jaipreet Virdi-Dhesi, 'Curtis' Cephaloscope: Deafness and the Making of Surgical Authority in London, 1816–1845,' *Bulletin for the History of Medicine* 87:3 (2013): 347–377.
21. Yearsley alludes to Graham Bell's piece 'The formation of a Deaf Variety of the Human Race' in both of his 1911 pieces discussed below, cited alternatively as 'On the Formation of a Deaf Variety of the Human Species in America.'
22. Kerr cites verbatim from MacLeod Yearsley's research in his report to the London County Council, 'The Deaf' *Annual Report Of The Council,* 3 (1910), 3: 160–165, especially 163.
23. Macleod Yearsley, 'The Education Of The Deaf: Its Present State In England, With Suggestions As To Its Future Modifications And Development.' Part III *The Lancet,* (March 1911): 652–657. Articles I. and II. were published in *The Lancet* of Feb. 25th 1911, 495 and March 4th, 574 respectively.
24. MacLeod Yearsley, 'Eugenics and Congenital Deaf-Mutism,' *Eugenics Review,* 2(4) (1911): 299–312, 311.
25. MacLeod Yearsley, 'Eugenics and Congenital Deaf-Mutism,' is discussed in G.R. Searle *Eugenics and Politics in Britain,* 93 and Alan H. Bittles *Consanguinity in Context* (Cambridge: Cambridge University Press, 2012), 55.
26. Yearsley listed these as Infective Diseases,—measles, scarlet fever, diphtheria, pertussis, influenza, German measles, enteric fever, meningitis, mumps and chicken-pox, pneumonia, rheumatic fever, tuberculosis and congenital syphilis, as well diseases of the ear and concussion (other miscellaneous causes including shock and lightning strikes). Macleod Yearsley, 'The Causes Leading To Educational Deafness In Children,' *The Lancet,* 180 (1912): 228–234.
27. Macleod Yearsley, 'The Causes Leading To Educational Deafness In Children, 234.
28. Helen Keller, *Helen Keller in Scotland/a personal record written by herself/ edited with an introduction by James Kerr Love,* 1933.
29. J(ames) Kerr Love, M.D. (Aural Surgeon, Glasgow Royal Infirmary, Aurist to Glasgow Institution to the Deaf and Dumb, Lecturer in Aural Surgery, St Mungo's College Glasgow, etc.), *The Causes and Prevention of Deafness: four lectures delivered under the auspices of the National Bureau for Promoting the General Welfare of the Deaf from lectures delivered 1912–1913* (London, National Bureau for Promoting the General Welfare of the Deaf: 1913). The audience and context of this lecture is discussed in Barrie H. Newton, 'The Plight of the Deaf in Britain, USA and Germany from 1880s to 1930s: a comparison of the social, educational and political links with the eugenic movements,' *The Galton Institute Newsletter,* Number 62 (March 2007): 2–8.

30. Kerr Love, *The Causes and Prevention of Deafness*, 14–15.
31. Kerr Love, *The Causes and Prevention of Deafness*, 5.
32. Kerr Love perhaps alludes here to the terms of the UK's recent Mental Deficiency Act of 1913 which treated a broad range of disabilities as warranting removal from society into institutions, with deaf people presuming to be part of a broader group of the 'feeble-minded'. See Anne Digby, 'Contexts and perspectives' in *From Idiocy to Mental Deficiency: Historical Perspectives on People With Learning Disabilities*, eds. David Wright and Anne Digby (London: Routledge, 1996): 1–21.
33. Kerr Love, *The Causes and Prevention of Deafness*, 14–15.
34. Kerr Love, *The Causes and Prevention of Deafness*, 14–15.
35. Kerr Love, *The Causes and Prevention of Deafness*, 2.
36. 'Our book shelf: The Causes and Prevention of Deafness. Dr. Kerr Love's Lectures' *British Deaf Times* in August 10(1913), 192.
37. Dan Kevles, *In the Name of Eugenics Genetics and the Uses of Human Heredity* (Cambridge MA: Harvard University Press, 1985).
38. On the medicalisation of deaf people see Harlan L. Lane, 'The Medicalisation of Cultural Deafness in Historical Perspective' in *Looking Back: A Reader on the History of Deaf Communities and Their Sign Languages*. ed. Renate Fischer and Harlan L. Lane. vol. 20, *International Studies on Sign Language and Communication of the Deaf* (Hamburg: Signum, 1993): 479–494.

CHAPTER 7

Institutionally Organizing for Hearing Loss

Abstract This looks at the organisations set up to advocate the rights of deaf and hard of hearing people in response to eugenic and economic discrimination against them. We focus first on the *Deaf and Dumb Times* as a forum for the heterogeneous deaf community with hard of hearing journalists articulating the nature of their exploitation and repression. Next we look at how the National Institute for the Deaf in 1924 emerged to a new umbrella role: one key factor was how the First World War changed the perceptions of deafness through new sympathy for combatant hearing loss. This transformed the advocacy of the pre-War organisations into a more unified national approach to defend the needs of hard of hearing people to trustworthy advice and support.

Keywords Deaf journals · First world war · National institute for the deaf

This final chapter looks at the organisations that emerged from the late nineteenth century to advocate the rights of deaf and hard of hearing people. First we look at how a series of periodicals that evolved from the *Deaf and Dumb Times* launched in 1889 (up to the *British Deaf Times* in the twentieth century) provided a forum for the heterogeneous deaf community marginalised by government initiatives. We will see how hard of hearing journalists such as George Frankland played significant roles in writing for these journals, especially in efforts to make them financially

G. Gooday and K. Sayer, *Managing the Experience of Hearing Loss in Britain, 1830–1930*, DOI 10.1057/978-1-137-40686-6_7

viable by broadening their appeal across the many deaf constituencies. Next we look at an alternative campaigning institution launched in 1911 as the 'National Bureau for the Encouragement of the Welfare of the Deaf.' Supported by the hard of hearing banker, Leo Bonn, this took the complementary aim of seeking to create employment opportunities for the deaf and hard of hearing. The Bureau struggled somewhat in this guise as employers were largely indifferent for reasons pertaining to insurance legislation and eugenic campaigns discussed in the previous chapter.

Nevertheless, the advent of the First World War changed much of the context for this. This was not just because there were more opportunities for Deaf people to be given work in noisy munitions factories intolerable to others. Of broader significance was the widespread experience of hearing loss at the battle front and indeed the home front by explosions at close range. This new mass cause of acquired deafness for servicemen and some civilian women generated greater awareness of and sympathy for adult hearing loss among both the public and the medical profession. After the war, attempts to get greater pension entitlements for the deafened ex-combatants was one factor that led to the relaunch of the Bureau in the new format of the National Institute for the Deaf in 1924. We shall see how far this organisation was able to respond to the post-WW1 imperative to actively assist the diverse deaf community with their strongly differentiated needs. Overall we will see how the context of combatant hearing loss in the First World War transformed the advocacy of the pre-War organisations into a more unified national approach that brought at least some genuine change to the circumstances of the hard of hearing.

Although overall this is a national story, it begins in Leeds. While there are many regional stories of the deaf and hard of hearing to be pursued, this is the city where developments in the last third of the nineteenth century helped to precipitate a new self-organisation among the broad community of deaf peoples.

1 Journals for the Broad 'Deaf' Community

Throughout the nineteenth century many local charitable organizations existed to support the work of the deaf and hard of hearing.[1] Yet there was little over-arching national co-ordination or monetary support for this until the later 1880s. As Peter Jackson notes, the National Deaf and Dumb Society, launched in 1879, collapsed in strife 7 years later.

Particularly pressing for such organisations were the disputes over the outcome of the Milan congress on deaf education in 1880 (see previous chapters). The UK government's Royal Commission on the Blind, Deaf and Dumb investigated the broader conditions of life for these constituencies between 1884 and 1888 and reported in 1889 without ever consulting them for their perspectives or recommendations. As Jackson observes, the outraged response to such a paternalistic snub, exacerbated by a controversial commitment to oralism, was to organise anew. One key development was the launch of the British Deaf and Dumb Association (later the British Deaf Association) which, in 1890, held its first congress at Leeds, a city which had hosted the United Institution for the Blind and the Deaf and Dumb since 1866.[2]

The city did not only serve as the obvious location for a national congress. In 1889 the monthly *Deaf and Dumb Times*, 'an organ intended for the Welfare of the Deaf and Dumb' was launched by two Yorkshiremen based in Leeds. These were Charles Gorham, Deaf son of a Masham vicar, and Joseph Hepworth born in Wakefield; having lost his hearing at age 8, Hepworth retained his speech until his death in 1921.[3] These two individuals were especially well placed to document the congress of the British Deaf and Dumb Association, of which Gorham himself was Secretary. Closely involved with all the successor journals to the *Deaf and Dumb Times*, Gorham and Hepworth included many categories of deafness within the remit of these publications. The changing titles, entangled with re-categorizations of deafness, were as follows:

The Deaf and Dumb Times (1889-91) published in Leeds

Deaf Chronicle (Nov.1891 to 1892) published in Leeds

British Deaf-Mute, and Deaf Chronicle no.13-48 (Nov.1892-Oct.1895)

British Deaf-Mute (Nov.1895-Oct.1896)

British Deaf Monthly: Organ of The National Association of Teachers of the Deaf, Etc. (1896-1903)

British Deaf Times: an illustrated monthly magazine for the deaf and dumb and those interested in their welfare. (1903-1954) published in Cardiff,

British Deaf News, Feb. 1956 - Sept. 2003.[4]

These periodicals supplied specialist journalism and editorial advice for the needs of all categories of deafness, as well as cut-and-paste extracts from other publications pertaining to deafness. They also furnished a forum for a broad spectrum of deaf readers to debate issues of the day. These journals indeed offered an umbrella vision, but struggled financially, taxonomically and indeed practically to get a sufficient readership to maintain the viability of their publications. The change of title from *The Deaf and Dumb Times* to the *Deaf Chronicle* in November 1891 was apparently prompted by Gorham's involvement in financial complications resulting in his bankruptcy, and a consequent need to close down the original publication. Hepworth relaunched with new editorial staff (Gorham still included) at his Leeds base. On consulting with subscribers and missionaries, the alternative title of *Silent Critic* was rejected, as was reference to the 'Dumb' (purportedly) to achieve economy of text.[5] Despite the flourishing awareness of Deaf schools and communities in the 1890s emphasised by Peter Jackson, the *Deaf Chronicle* still struggled to find a readership, merging with the *British Deaf-Mute* a year later.

In the meantime, we can see in the pages of the *Deaf Chronicle* how it represented all conditions of deafness, including the hard of hearing and their travails with exploitative 'cure' merchants. This was the era of the new journalism in which writers for newspapers and magazines did not passively report on the world around them, but sought actively to expose crime and fraud.[6] Thus in the Capanbells column, readers regularly saw its campaign against the 'Quack Doctors who profess power to cure deafness': the column warned readers not to believe newspaper advertisements declaring 'Deafness Curable' with new ear gadgets. To supplement this in 1892 it reproduced in its entirety a piece concerning 'Swindles on Deaf People' recently published in *Tit-Bits* by a 'partially deaf' journalist (See Chap. 5).

Nevertheless, these merchants of mock-cures still advertised undeterred by the time that the *Deaf Chronicle* had evolved again into the *British Deaf-Mute*, in late 1895. Such was the relentlessness of their advertising campaign, that in 1895, the partially-deaf house journalist George Frankland wrote a comparative piece titled 'Aids to deafness.' Comparing the treatments of the quasi-medical 'Aurists', the high-street 'Auricians' and the Newspaper-advertising 'Quacks' he wrote:

> Aurists have syringed, painted, oiled, physicked, inflated and perforated me. Auricians have furnished me with diaphragms, trumpets, whispering

> tubes and noise machines. Quacks have sent me their works, exhibited their devices, and endeavoured to bleed me. So, by this time, I ought to be an authority on any subject. The general result of my experience has been to bias me in favour of the regular aurists and auricians.[7]

As Frankland explained further the respectable aurists and auricians were to be trusted because they assiduously kept 'abreast of the latest scientific discoveries' and thus were more likely to capitalize upon innovations than were the 'untrained amateurs' to whom the advertised devices were commonly due. Lastly, as the regular practitioners did not have to 'advertise very largely', they could afford 'to sell their goods at a moderate profit'—not at the exorbitant prices demanded by the 'Quacks'.[8]

One 'persistent' advertiser in this category was a Dr. J. H. Nicholson. At a cost of two guineas, the artificial 'Ear Drums' sold by Nicholson were too costly for Frankland to purchase; yet Nicholson nevertheless kept bombarding him with pamphlets about them. Eventually, at a surgical instrument shop, Frankland obtained a similar appliance—a rubber disc attached to a wire—at a twentieth of Nicholson's price but found it had no therapeutic value. Another energetic advertiser Frankland exposed was the 'plausible, bustling' Rev. Silverton whom Frankland had seen in Liverpool bearing 'all manner of shining and expensive serpent tubes and trumpets'. Frankland's greatest objection to Silverton was not just the ineffectiveness of any of the items that he peddled, but his complete lack of professional ethics—a conspicuous characteristic he alleged of all hearing aid 'advertisers'. Whereas at a 'respectable aural establishment' it was easy to secure a hearing aid on trial with cash returned if useless; but this was 'not the practice of our advertising friends'. The key issue that Frankland discerned was commercial success not assistance to the deafened. 'Perchance', he archly remarked, a sale-or-return policy 'would not be profitable'.[9]

Columnists for the *British Deaf-Mute* and its successor *The British Deaf Times* regularly warned its readers against opportunists like Nicholson and Silverton. Such journals, did, however, welcome and endorse new electrical gadgets that came along at the turn of the century, based on the microphone amplification technologies of the telephone—and later the amplifying valves of early wireless (radio) sets. Thus, for example, in 1911, the Globe Ear-phone imported from the USA was the subject of a glowing review in *The British Deaf Times*, contrasted favourably with recent products by the American Miller Reece

Hutchison (the 'Akoulallion' and patented 'Acousticon'). The operations of this device were clearly explained in the accompanying literature, and trials at home with money-back guarantee were offered: these two characteristics were soon to become standard features of the trust relationship between vendor and makers of hearing aids. More than this, readers of the journal were asked to become active experimenters and commentators on the merits of these devices.[10]

However, in the context of political change, especially to greater opportunities for the various groups of deaf people, to laws concerning fraudulent advertising and vendor liability, the journals largely reflected the diverse views of their readers without engaging strongly with the wider hearing community—including government. Equally the journals were not well placed to bring unity of purpose to the multiple groups representing the variety of deaf interests. It was in that vein that a new national bureau was launched in 1911, seeking to change the overall position of the broad deaf community in an increasingly challenging world.

2 The National Bureau

Complementing the activist role of the journals discussed above but getting little acknowledgement from them at first was the National Bureau for the Encouragement of the Welfare of the Deaf. Certainly the Bureau's initial goal was not to compete with the *British Deaf Times* in tackling fraudulent vendors of hearing aids or debating the appropriate education in oralism or signing. Rather its overall aims were to publicize more broadly the various challenges facing people in Britain experiencing deafness, and coordinate their various representative bodies, whether local or national. This appears to have been a movement to coordinate that was without precedent in other countries—no similar organisations emerged in the USA until after World War 2.[11]

The key figure in this was the wealthy German-born Jewish banker Leo Bonn. In his sixtieth year he had encountered the lip-reading teacher Mary Hare who both offered him lessons and cultivated his broader interest in issues pertaining to hearing loss and deafness:[12]

> as I found my hearing getting worse I asked Miss Hare, a teacher of lip reading, to give me some lessons, she interested me in the cause of the deaf and after examining the question with the help of Mr. Story, head of

> a big school for the deaf at Stoke-on-Trent, who moved me by his enthusiasm, I promised to devote a certain sum of money to starting a clearing house to look after the interests of the deaf.[13]

Following such experiences Bonn not only financed but personally initiated the National Bureau in 1911–1912. The first meeting was held in the dining room of his own home with 100 individuals concerned with the various kinds and categories of deafness. He took the chair with about one hundred attendees interested in general deaf welfare including Lord Fletcher Moulton, Lord George Hamilton, Sir Frederick Milner, and Lady O'Hagan. The Minute book of the Bureau's first meeting on 9th June 1911 shows that among those present were both James Kerr Love and MacLeod Yearsley—a broad church indeed. Through such patronage Bonn's aim was not to introduce lip-reading oralist methods to all who experienced deafness. His was a much more latitudinarian goal of acting as a coordinating point for the various groups involved in deafness so as get the broader public interested in and supportive of the various issues of deafness. As he later put it 'I felt I had to be the rallying point, for there were many different interests to be reconciled'.[14] In addressing the Bureau he declared that his 'own case of bad hearing' offered an 'excuse for his venture'; moreover, his 'strong feeling for fellow-sufferers, perhaps less happily placed in life', who had generally 'urged him to take action' for many kinds of deaf people.[15]

Bonn proposed three main objects for the Bureau—centralization: to communicate and cooperate with all relevant deaf bodies; information gathering from government, periodicals, and daily papers, and investigation of 'any problem affecting the deaf'—with results submitted to its Council for private and public 'propaganda'.[16] In December 1912 Bonn gave the Bureau an endowment of £5000 to fund it for its next 5 years, working also to raise subscriptions. In a more activist vein, he compiled a directory of deaf organizations, and sought Home Office co-operation to promote employment opportunities. This was a major concern given the 1911 Employers Liability Act and Insurance Act that obliged employers to apply medical tests to (prospective) employees that might reveal evidence (among other things) of hearing loss.[17]

In the face of indifference or outright hostility to the promotion of employment for all categories of deaf and hard of hearing people, it is revealing that Bonn's public strategy was not to win sympathy for these constituencies so much as to plan to eliminate them, echoing wider eugenic themes discussed in the previous chapter. Bonn thus commissioned James Kerr Love's four lectures on *The Causes and Prevention of Deafness* (1913) and concurrently campaigned to ensure that the eugenically motivated Mental Deficiency Act (1913), was not illegitimately applied to send deaf people (e.g. misunderstood as 'Feeble-minded persons') into the proposed new institutions or 'colonies' of the 'deficient'. [18]

Yet even as this organization sought to nurture the well-being of the hard of hearing and deaf, the exigencies of the First World War created a new challenge. As was noted at the Bureau's third meeting in November 1914, just three months after the conflict broke out:

> The medical aspects of deafness, ably dealt with by the Medical Committee, were now brought to the front in a somewhat unusual way. The heavy percussion of modern artillery was likely to affect hearing, and they had felt themselves called upon to offer the services of the Bureau to the War Office and the Admiralty, in order to assist them in dealing with soldiers and sailors who may suffer from deafness in consequence of the war.[19]

While prescient of the widespread deafness that would result from a prolonged war, it was not, however, the Bureau that would come to the aid of those who lost part or all of their hearing in the coming conflict.

3 The Rise of the War-Deafened

As noted in the previous chapter, the First World War created a new realm of opportunity for physicians to show how they could diagnose and treat hearing loss specifically for adults—rather than the constituency of children assumed in Kerr Love's educational treatise. As discussions of 'war-deafness' in *The Lancet* show between 1916 and 1918, it is clear that a new kind of medical expertise was developing largely unconnected to eugenics. Deafened combatants were not treated as being genetically predisposed to loss of hearing, but the honourable victims of the damaging percussive effects of artillery.[20] The physicians associated with Bonn's Bureau were not, however, directly involved in this new project of studying war deafness.

This greater visibility of the war-deafened was noted by others, not least Helen Keller, as an opportunity to promote the broader cause. As she wrote to James Kerr Love in November 1916 as the war entered its third year and the USA pondered whether to join it: 'I think sympathy aroused by the funds for relief of the soldiers deafened and blinded in the war is reacting to a certain extent to benefit our own deaf and blind.'[21] But this did not seem to be the case for the Bureau which withered in its operations as so many resources and personnel were directed to other matters. When the British Government's Pensions minister set up an Aural Board in October 1917 to deal with cases of soldiers discharged through acquired deafness, it was not to the associates of the ailing Leon Bonn that it turned for advice, but Major Dundas Grant, consulting surgeon to the Central London, Nose, Throat and Ear Hospital. Under Dundas Grant's chairing, the Aural Board offered classes in lip-reading to the discharged servicemen at the College of the National Association for the Oral Instruction of the Deaf in Fitzroy Square.[22]

This Government's 'one-size-fits-all' policy of offering lip-reading classes was probably calculated to be the cheapest and easiest plan to implement. But it was by no means what the deafened servicemen wanted. Indeed it was the periodical press rather than Bonn's Bureau that highlighted the disaffection of the deafened servicemen with this single offer. In the summer of 1918 *The British Deaf Times* took up the cause of the serviceman partially or totally deafened by war insisting that discretionary and bespoke solutions were required to assimilate them back into ordinary working life. It complained that some of the 'deafened lads' just wanted their pre-war job back (or some equivalent) with a 'war bounty' and then be left to carry it on without any obligation for hearing. Others, it argued, should be provided with new kinds of employment in which hearing would not be required. The assumption that lip-reading was sufficient for most cases was dismissed by Hepworth and his editorial team: as far as they were concerned that practice was fit only for the 'home circle' or with 'good friends.' In the practical environment of commerce the attempt to 'read anyone's speech with the eye' was not only practically difficult except in favourable circumstances. Much worse, some hearing business people behaved disingenuously, back-tracking on apparent verbal agreements that had been lip-read, claiming that the lip reader had misunderstood what had been uttered.[23]

Such were the apparently anti-oralist commitments of *The British Deaf Times* that it did not campaign for the Government to issue deafened soldiers with hearing aids, notwithstanding previous recommendations concerning new electrical gadgets which might assist the partially deaf in recovering lost hearing (See Chap. 4). However, we know that at least one such veteran did wish for just such a device in the war's aftermath: his request was acknowledged in the *British Deaf Times*. The correspondent, fashioning himself as 'One of the Victims' of the First World War had apparently appealed through the *Spectator*'s columns in late 1919 to the 'medical advisers' among its readership:

> A short-sighted man may go to his doctor and through him to an optician, and so get a certified instrument to remedy his defect. No such advantages are open to the partially deaf. The electric aids, admirable in conception and an inestimable boon, are…[often] carelessly made, supplied with difficulty, and without any guarantees, but their price is prohibitive to poor patients. This class has, unhappily, very greatly increased owing to the war. Surely the medical profession, when unable to cure this most disabling of infirmities, might recognize the mechanical aids in existence and help their patients to get them, properly made and at a reasonable price.[24]

One such electrical aid that might have prompted this concern was the Stols Electrophone advertised in the London *Times* during the Great War.[25] Such an appeal to expensive technological solutions was not immediately met by great interest or sympathy. The large majority of Government pensions to ex-servicemen was dedicated to those who had lost limbs, eyes, or been facially disfigured. While total hearing loss was the subject of compensation, no allowance was made for any hearing loss. This instead was treated by the National Benevolent Society for the Deaf and Deafened Ex-Service Men's Fund from 1919 to support soldiers in securing pensions, training, medical treatment, assistive instruments, etc.[26] Around 33,000 deafened ex-Service men received pensions from this fund after the Great War having no recourse to government funding dedicated more to blinded, disfigured and amputee veterans.[27]

Many of the deafened veterans were military officers who moved back to desk-based positions after the war. For those dissatisfied with lip-reading classes, there was a new opportunity for technological solution to return. One electronics company that sought commercial work in producing an effective valve-amplifying desk hearing aid for the work place was Marconi. This global company had produced many of the wireless radio sets used by the military services during the war, using its lucrative thermionic valve patent.[28] Once this patent expired in 1918, and the profitable war period was over, the Marconi Company sought new markets, one of which was the now increased number of servicemen who had lost their hearing.

Recognizing the new higher profile of physicians in mitigating deafness (now no longer just seeking to prevent it) the Marconi Company approached the London otologist William Mayhew Mollison. As a leading figure in otology, he could supply them with formal clinical credentials the company needed to promote its new hearing aid to middle class audiences. In May 1923 Mollison presented the Otophone to the Otology section of the Royal Society of Medicine as an 'instrument for assisting the deaf.' Mollison commented that after only a week's experience of the instrument using amplifying valves already he had found it 'useful in helping really deaf people to hear', devices based on telephonic microphones, having been much less effective. Nevertheless, observing the huge bulk of the device taken up with heavy electrical valves and circuitry, the Section chairman Sir Charles A. Balance noted less flatteringly that 'a great obstacle to its use was its weight'. Tis could not be used as a portable device for everyday interaction by a deafened serviceman. Moreover, the price of the Otophone was not discussed: as large and complicated device, it was targeted only at the wealthiest deafened subject with scope for a dedicated desk hearing aid in their workplace or domestic office.[29]

What later became the Marconiphone company planned initially not to market this gigantic instrument through the 'usual trade channels', proposing to supply it from its 'research laboratory' under the recommendation of the medical profession. A decade later, however, this device was being marketed by the Hawksley company which advertised it in the London *Times* newspaper.[30] Yet the medical men and hearing aid vendor did not have it all their own way in harnessing the new practical sympathy for acquired deafness wrought by the horrors of the First World War. By the time the Otophone was launched upon the world in 1924, the larger community of deaf peoples had its own agency to assert. With the patronage of senior politicians, leaders of the effectively defunct Bureau had negotiated what was effectively a merger with other deaf organisations, and the National Institute for the Deaf was born.

4 The National Institute for the Deaf

In a world recovering from the First World War, support for the broader issues of disabilities was not given the highest financial priority by government or private organisations. But in 1924 new developments emerged. As Peter Jackson has noted, sport among deaf communities flourished in the interwar period, a major public leisure activity which allowed for multiple forms of communication.[31] At the global level, the First International Games for the Deaf (originally known also as the First International Silent Games, now the Deaflympics) was held in Paris in 1924, 2 weeks after the Olympic Games. Nine European countries participated, sending deaf athletes (mostly male): Belgium, France, Britain, Hungary, Italy, Latvia, Netherlands, Poland, and Romania. This of itself signalled a new international coordination among the deaf communities at least at the level of sport.[32]

However, in Britain the emergence of the National Institute for the Deaf (NID) in 1924 emerged not from international imperatives as such, but rather from the alliance of multiple regional organisations for the hard of hearing within the UK negotiated by the vestigial membership of Leo Bonn's Bureau. Characteristically of charities for the disabled, the NID was headed in its first 11 years by an aristocratic patron. This was the fully hearing Godfrey Rathbone Benson, an Oxford Philosophy Don and Liberal who became a supporter of many national charities when elevated to the House of Lords as Lord Charnwood of Castle Donington in 1911.[33]

At the opening conference on March 19th 1924 for the newly relaunched NID, Baron Charnwood's opening address revealed the hearing person's stereotyped emotional responses to communicating with the group that he represented:

> I have two things that ought to be said about the deaf. Their misfortune is not one that instantly appeals to sympathy. Everyone sympathizes with the blind. You do not instantly discover that the deaf are deaf. I confess frankly myself that my inclination is to be irritated with the person who is deaf.

Evidently having overcome this irritation enough to research the concerns of the broad deaf community, Charnwood's appeal to fund the co-ordinated teaching and care of deaf children was grounded instrumentally in the Oralist economics of employment. Appealing to arguments familiar from 19th century debates on employment and Oralism, he argued that with such support the 'deaf could become the most economically remunerative of all the classes known as special'. Such supportive work he declared was 'the only hope for the deaf':

> Its importance to the nation is apparent when it is remembered that there are some 40,000 deaf and dumb persons within its borders, not to mention a vastly larger army of others who are partially deaf or hard of hearing, and who thus need special help to render life endurable'.[34]

While much of the NID's labours were dedicated to the most needy and under-employed 'deaf', it also dedicated itself to supporting that much 'larger army' of the hard of hearing. Within a few years it had secured for itself the role of policing the vending of hearing aids from companies less reputable than Hawksley or Marconiphone.

By 1929, the 5 year old National Institute for the Deaf had seen enough of the predatory behaviour of certain commercial operations, especially when extended to the deafened community created by the Great War. With its new found representative status for all kinds of hard of hearing people it decided to act as a politically activist organisation. It took the radical step of launching an easily-requested register of firms and dealers who would make no unscheduled house calls, offer disinterested advice on the suitability of any electrical or mechanical device; and offer a full refund if any device proved unsatisfactory.

> Advertisements, encouraging the deafened, regardless of the nature or degree of their auditory defect, to expect the return of normal hearing, prey to-day, more than ever before, upon their natural hope for relief; and large numbers of hearing aids are purchased only to be cast aside as useless.

Certain dealers simply refused to allow any 'adequate trial' of their instruments before purchase or to offer refund for ineffective purchases. Such advertisers should thus be 'compelled to adjust their misleading advertisements' on the nature of deafness and the 'possible performance of their instruments', and indeed to 'amend their methods of business to ensure a fair deal to the deafened'. The NID's committee had thus devoted considerable attention to the best method of 'safeguarding the deafened' against the disappointments and financial hardships entailed by such firms that it archly described as only being concerned for deafness to 'the extent of their sales'. The NID thus launched a register of 'reliable firms and dealers who conduct their business on these fair conditions': these were committed to making no unscheduled house calls; offer only disinterested advice on the suitability of any electrical or mechanical device; and offer a full refund if any device proved unsatisfactory. 'Deafened persons' were strongly advised to deal only with those who met these conditions and could receive a copy this list from the NID by submitting the cost of postage.[35]

It was not until 1950, when the advent of the NHS Medresco hearing aid had put the private hearing aid manufacturers onto the back foot, that the NID published its list of approved dealers for the cities and major towns in the UK and Ireland. Many older names such as Bell, Down, Hawksley and Rein can be found there still selling mechanical aids such as trumpets, speaking tubes and auricles, along with other interwar companies: Amplivox, Multitone, Oticon and Philips that represented the enormous growth of the hearing aid industry when electronic amplifying valve technology became available in the interwar period. By this time all dealers, of whatever kinds of device needed a visible NID certificate to offer merchandise under its authentication. Importantly, however, the NID also indicated in this same document that lip-reading skills were advisable for 'deafened persons' whether they used a hearing aid or not, and it offered recommendations for books on lip-reading to this end. Yet even as new university experts such as Irene Ewing promoted new electronic hearing aids as necessary support for effective lip-reading, nevertheless, still many preferred to use an acoustic hearing device modelled on Harriet Martineau's in the new National Health Service of 1948.[36] Thus did hard of hearing people assert their discretion to select historically grounded techniques for communicating, whatever the NID might recommend for modern electronic hearing aids.

5 Conclusion

In this chapter, and the one before, we have seen that organised concern for the welfare of all categories of deafness (including hearing loss) was prompted by their unsympathetic treatment from the British state—as well as employers—in the last two decades of the nineteenth century. This was for reasons linked to the economics of employment, the eugenics of eliminating people with non-standardised capacities and the rise of an oralist culture that normalised speech/hearing and pathologised sign language. Finding it harder to get political recognition and practical support for their situation, the deaf journals and the National Bureau sought to get public recognition of such people's various claims—whether for deafness per se or for loss of hearing. It is unclear to what extent they succeeded before the Great War, but certainly at least some the claims of deafened servicemen who had lost hearing through combat conditions were eventually formally acknowledged. It is evident, however, that not all were satisfied with the options offered them—mostly lip-reading classes and pensions at a lower level than for blinded servicemen. This was a long way from meeting the requests from deafened war veterans for either jobs for the deaf or easy access to reliable hearing aids. It was such disaffection that prompted the rise of the National Institute for the Deaf in the interwar period, as well as the new regimes of hearing aid supply for adults that emerged in the 1930s.

Notes

1. Martin Atherton, *Deafness, community and culture in Britain: Leisure and cohesion, 1945–1995* (Manchester: Manchester University Press, 2012), 33–47 and Peter Jackson, *Britain's Deaf Heritage* (Edinburgh: Pentland Press Limited, 1990), 37–158.
2. Jackson, *Britain's Deaf Heritage*, 115, 382.
3. On Hepworth see Jackson, *Britain's Deaf Heritage,* 145–146.
4. Amanda Nichola Bergen observes that the importance of the Yorkshire community, and Leeds in particular, was epitomized by the choice of Leeds as venue for the first congress of the British Deaf and Dumb Association in July 1890, *The Blind, the Deaf and the Halt: Physical Disability, the Poor Law and Charity c. 1830–1890, with particular reference to the County of Yorkshire* (University of Leeds: unpublished Ph.D. thesis, 2004) 400. Brian Grant, *The Deaf Advance: a history of the British Deaf Association, 1890–1990* (Edinburgh: Pentland Press 1990), 21. See Dominic Stiles' blog at http://blogs.ucl.ac.uk/library-rnid/2013/12/13/joseph-hepworth-i-thought-that-there-were-probably-only-about-a-dozen-deaf-men-in-the-wide-world/ (last accessed 14th Dec 2016).
5. 'Hepworth wrote 'Our correspondents, authors, as well as article writers, are however, welcome to use the words 'and dumb' if they wish, but we might give the hint that it would save our space considerably if these were omitted as much as possible.' All communications, news,

and subscriptions were to be directed to Messrs Hepworth & Lunn, 16 Howarth Place, Camp Road, Leeds. 'Editorial Chat: How is it?' *The Deaf Chronicle* 1 (1891) 1.

6. On the trend of a fearless 'New Journalism' for exposing crime and exploitation in the 1890s, see Gowan Dawson, 'The Review of Reviews and the new journalism in late Victorian Britain', in *Science in the Nineteenth-Century Periodical: Reading the Magazine of Nature*, eds. Geoffrey Cantor and Sally Shuttleworth (Cambridge: Cambridge University Press, 2004): 172–195.
7. George Frankland, 'Aids to Deafness,' *British Deaf-Mute*, 5 (1895), 84.
8. Frankland, 'Aids to Deafness,' 84.
9. Frankland, 'Aids to Deafness,' 84.
10. 'The Globe Ear-phone', *British Deaf Times* 8 (1911): 127–128.
11. Kenneth W. Berger, *The Hearing Aid: its operation and development* (Michigan: USA National Hearing Aid Society, 1974), revised edition, 151–160.
12. Jennifer Little, *Mary Adelaide Hare 1865–1945*, 1983. http://www.maryharehistory.org.uk/articles/staff_mary_hare.html (16th Oct 2016).
13. For more on Bonn's role in the origins of the Bureau, see Leo Bonn, *Unpublished Memoirs*, 19 held in the Action on Hearing Loss library (thanks to Dominic Stiles for this). Action on Hearing Loss, *Celebrating 100 years of Action on Hearing Loss* (London: Action on Hearing Loss, 2011), and http://blogs.ucl.ac.uk/library-rnid/2012/04/13/leo-bonn-founder-of-the-national-bureau-for-promoting-the-general-welfare-of-the-deaf/ (14th Oct 2016). Story's School was the Mount Blind and Deaf School in Stoke on Trent.
14. Bonn, *Unpublished Memoirs*; Royal National Institute for the Deaf, *RNID, 1911–1971: sixty years of service to the deaf* (1971); B. Grant, *The Deaf Advance: a history of the British Deaf Association, 1890–1990* (1990) Anne Pimlott Baker, 'Leon Bonn' *Oxford Dictionary of National Biography*, http://0-www.oxforddnb.com.wam.leeds.ac.uk/view/article/68991 (viewed 20th September 2016).
15. *Minute book of the National Bureau for the Encouragement of the Welfare of the Deaf*, Action on Hearing Loss Library, 1.
16. *Minute book of the National Bureau,* 3. Barrie Newton focuses on the aims of the *National Bureau, First Annual Report*, (1912) was '… to benefit the sufferers, to render their lives happier and more endurable and to prevent them in any way becoming a hindrance to the progress of society'. Barrie H. Newton, 'The Plight of the Deaf in Britain, USA and Germany from 1880s to 1930s, 2–8.
17. *Minute book of the National Bureau* 2nd report November 12, 1912, 56–58.

18. Among the physicians that the Bureau brought on as its advisors, James Kerr Love was included, but MacLeod Yearsley was not.
19. 'Report of the Third Annual Meeting, Nov. 11, 1914,' *Minute Book* of *the National Bureau* 6 AoHL Library.
20. John F. O'Malley, 'Warfare Neuroses of The Throat And Ear.', *The Lancet*, May 27, 1916, 1080–1082; 'War Injuries And Neuroses Of The Ear. *The Lancet*, Feb 24, 1917, 304–305; 'War Deafnesses [Editorial].' *The Lancet*, October 13, 1917, 576; P. Mcbride and A. Logan Turner War Deafness, With Special Reference to The Value of Vestibular Tests, *The Lancet*, July 20, 1918, 73–74. In his final works after World War 1, W.H. Pitt Rivers also noted that deafness was one of a range of symptoms including 'paralysis, mutism, contracture, [and] blindness' that had been be caused by 'war neurosis'. W.H. Pitt Rivers *Instinct and the Unconscious: A Contribution to A Biological Theory of the Psycho-Neuroses* (Cambridge: Cambridge University Press, 1922), 128, 129, 206.
21. Letter from Helen Keller to James Kerr Love, Sept 26, 1916, in James Kerr Love (Editor) Helen *Keller in Scotland: A Personal Record Written by Herself* (London: Methuen 1933), 73. The British Deaf Times contained numerous discussions on this point 1916–1919.
22. 'Care of the Wounded', *British Journal of Nursing*, October 13, 1917, 238.
23. 'Chat with our readers by the Editor: "The Deaf Soldier", "What Deaf Soldiers Want",' *British Deaf Times,* July—August 1918, 56.
24. 'One of the Victims', [undated letter] 'Deafness', *British Deaf Times* 15 (1919) 4. The editorial claims that this letter was 'culled' from *The Spectator*, but it has not been possible to locate this letter in that latter journal.
25. E.g. Stols electrophone advertisement from *the Times*, Wednesday, July 05, 1916, 3; this company is at the same address as the British Stolz company was at in 1915 'The Stolz Electrophone Co' The Times, Nov 09, 1915; 3. Originally a US company based in Chicago, the British branch went bankrupt c. 1914–1915, and evidently changed its name in Stols in 1916 to sound less Germanic once conscription for military service had commenced.
26. Minutes from the Deafened Ex-Service Men's Fund, Action on Hearing Loss Library. Thanks to Coreen McGuire for pointing us to this source.
27. R. Scott Stevenson, 'The Otologist and Rehabilitation of the Deaf (President's Address)' *Proceedings of the Royal Society of Medicine* Vol. 42 (1948), 55.
28. Graeme Gooday, 'Combative Patenting: military entrepreneurship in First World War telecommunications,' *Studies in History and Philosophy of Science—A,* Vol 44 (2), June 2013: 247–258.

29. 'Societies' Proceedings Royal Society of Medicine—Section of Otology' *Proceedings of the Royal Society of Medicine* May 1923, *Journal of Laryngology & Otology* Vol 38 (1923), 529.
30. 'T. Hawksley Ltd'. *The Times* 7 Feb. 1933: 14. Thanks to Jamie Stark for tracing this advertisement.
31. Jackson, *Britain's Deaf Heritage*, 222–226.
32. https://www.deaflympics.com/ (last accessed 16th October 2016).
33. Humphrey Sumner, 'Benson, Godfrey Rathbone, first Baron Charnwood (1864–1945)', rev. Marc Brodie, *Oxford Dictionary of National Biography*, Oxford University Press, 2004; online edn, Sept 2014 [http://0-www.oxforddnb.com.wam.leeds.ac.uk/view/article/30715, accessed 12 Dec 2016] Note that the British Deaf and Dumb Association continued entirely independently of the National Institute for the Deaf and led by David Fyfe, a Deaf activist from Glasgow who had risen from the position of apprentice brass-finisher. Jackson, *Britain's Deaf Heritage*, 216.
34. [Chairman's Address] The *National Institute for the Deaf: Report of Conference of Schools and Agencies and Persons interested in the Deaf, Held at the Kingsway Hall, London* (1924): 1–3.
35. 'Aids to Hearing,' *National Institute for the Deaf Annual Report,* (fifth edition) (1929): 15–16. This list of approved hearing aid vendors was not published openly until a decade later, presumably to avoid litigation from those companies that were not included.
36. Irene Ewing, *Lip-Reading and Hearing Aids* (Manchester: Manchester University Press, 1946). The Thackray Museum in Leeds contains a blueprint for an NHS version of a Martineau hearing aid from the 1970s. Thackray Medical Museum (TMM), Leeds, 1333.022, Trumpet hearing aid, telescopic cone shape, black celluloid, ear piece: vulcanite, c. 1890s–1900; National Health Service, Model OL370, blueprint for hearing apparatus, telescopic ear trumpet with bent end, celluloid, 1976.

Epilogue

This volume has explored hearing loss as a largely invisible domain of experience between the very different worlds of Hearing and Deafness. In showing how to interpret the changing circumstances and management of hearing loss we have rendered this domain more visible at least for the period 1830–1930. Yet what we have accomplished is largely a programmatic sketch of diverse trends and experiences. Much further research is now needed to recover more fully the lives of hard of hearing people. In this closing section, we point to future areas of investigation in the hope that readers will find some interesting opportunities to explore.

Recovering the lived experiences of hearing loss from the past has undeniably been difficult and somewhat serendipitous. In only a few cases have we found direct self-declared testimony from named hard of hearing individuals: the sociologist and professional writer, Harriet Martineau, the Rev. F.J. Hammond featured in our epigraph, and the campaigning journalist George Frankland. Among others less willing to expose their identity were an anonymous 'Victim' of the First World War demanding an electric hearing aid; the pseudonymous critic of hearing aid vendors, 'Evan Yellon,' and the isolated lonely women seeking mutual contacts via *The Woman At Home* corresponding under codenames such as 'Out in the Cold' and 'Depressed One.' Our hope is that from archival correspondence sources many more voices can be

G. Gooday and K. Sayer, *Managing the Experience of Hearing Loss in Britain, 1830–1930*, DOI 10.1057/978-1-137-40686-6

found that will give a broader demographic base– in terms of age, class, gender—to the evidential base of studies of past hard of hearing cultures.

One obvious source of evidence would be the Hard of Hearing clubs that emerged across the UK from the 1920s. These arose as otologists (especially James Kerr Love) developed their new prerogative for assisting hard of hearing people to socialise with each other and experiment with techniques in both using new electronic hearing aids and or lip-reading. Archival records and publications of these regional and civic clubs held in the Action on Hearing Loss library in London could be of great benefit in capturing the kinds of nucleus that were available for socialization, albeit under a medicalised umbrella organisation.

Of course, Hard of Hearing clubs only represented a minority of that constituency, specifically those willing publicly to self-identify as hard of hearing and to collaborate with otologists on developing technocratic solutions based in hearing aids and or the virtuosic subtleties of lip-reading. What is also needed, for example, is a deeper study of other organisational units run on different practical and institutional premises. A study of the early years of the campaigning National Institute of the Deaf could help us understand to what extent its membership was constituted by hard of hearing people as well as other kinds of self-identifying deaf people in strategic yet potentially fragile alliances.

In understanding that history we need to consider how far the NID managed the competing advice on the use of lip-reading and hearing aids as well as protecting its members from unscrupulous or fraudulent practitioners in both domains. We can similarly look for a better understanding of the agency of the NID's own membership in campaigning for improved facilities and how far relied upon the patronage of senior public figures to leverage change, such as access to subsidised hearing aids for First World War veterans. One place to look is in the House of Lords, for individuals such as Lord Leverhulme, who promoted the cause of the hard of hearing, in part through their own experiences of auditory loss in later life.

A less institutionally-focused approach would be to look for the experiences of hearing aid users as captured in the adaptations and wear and tear of the artefacts that they left behind. Many such are in private collections, but numerous models are now captured in museums. A more rigorous survey of collections of these artefacts will enable us to capture something of the daily symbiosis of hearing aid users and their devices from their markings and accompanying texts and provenance information. In seeking to understand the lives of the deaf, it would also be important to look at the rise of electronic hearing aids in the 1920s, and recognise the impact of the associated rise of the radio as a major means of communication in the 1930s. The Radio, just as much as the telephone had done in the preceeding decades, came to exclude and marginalise the hard of hearing in a Hearing world. Exploring the history of the silent/captioned film era, in comparison to the 'talkies', is also germaine here.

Moreover, from looking at the extraordinary array of such devices across national and international museums we will perhaps get a sense of how over their life course individuals changed their hearing aids for various reasons or perhaps even (among the wealthier) multiplied their ownership of them for different social contexts. Conversely, the records of the UK's Association of Hearing Aid Manufacturers could reveal much about how the hearing aid industry learned to respond to the needs of the hard of hearing population in developing devices better suited to their concerns, and also self-regulating to resist the reputational damage caused them by the more opportunist manufacturers.

The continued growth and refinement of training in lip-reading for hard of hearing adults, especially by women, could usefully focus on the educational work of such key individuals as Irene Goldsack (later Ewing) at the University of Manchester in the interwar period. As a hard of hearing individual appointed to teach controversial oralist methods in the first University appointment in the UK, she also was among the leaders of the movement to combine the hitherto competing elements of lip-reading to support the use of hearing aids. Her collaborations with student then spouse Alexander Ewing with the Leverhulme Trust and the Medical Research Council, would reveal how institutionalized research into support for hard of hearing people became managed at least in part by hard of hearing people themselves. Other leading women, such as audiometry specialist Phyllis Kerridge, at UCL—with whom the Ewings did not collaborate—can be analysed to ensure a balanced gendered account as a technocracy of Post Office male engineers worked with the Ewings to develop plans for standardised national hearing aids.

In the development of plans for standardised hearing aids, we need moreover a broader understanding of mid-20thC growth of otology as a discipline increasingly willing to collaborate with technologists. One example of this is the development of the 'Medresco', a standardised utility hearing aid designed for the hundreds of thousands unable to afford their own and made available via the UK's new National Health Service in 1948, albeit with many stories of fallbility and dissatisfaction. Elswhere this rise of state-sponsored technocratic support is especially important in the USA after World War 2 when the new specialist subdiscipline of audiology emerged to deal with many thousands of deafened servicemen, especially from the US air force. While audiology was not formally organised in Britain until the 1960s, this episode indicates how much comparative developments in other countries are essential to study: while lip-reading was conducted along local/national lines, professional and technological developments relating to hearing loss moved across national boundaries as much expertise was shared in the interwar period and after.

Finally, another comparative study needed is the broader investigation of how visual loss and hearing loss interacted, yet also with contrasting experiences. Very often we see the wearing of spectacles to compensate for myopia as much more mainstream, affordable and unproblematic than the wearing of hearing aids. More than that, what has emerged in the twenty-first century as the (ironic) fashionable visibility of lost visual capacity in glasses contrasts with the normative invisibility of hearing loss in lip-reading and minituarised hearing aids. Exploring the history of these and similar comparisons, and the associated cultures of humour and stigma, will enable future scholars to map the boundary issues that have often served to obscure the hard of hearing as 'other' in the historical record, leaving them with a history that has too often been neglected but can now be told.

Index

G. Gooday and K. Sayer, *Managing the Experience of Hearing Loss in Britain, 1830–1930*, DOI 10.1057/978-1-137-40686-6